Targeting Biofilms in Translational Research, Device Development, and Industrial Sectors

Dustin L. Williams
Editor

Targeting Biofilms in Translational Research, Device Development, and Industrial Sectors

Springer

Editor
Dustin L. Williams
Department of Veterans Affairs
Salt Lake City, UT, USA

Department of Orthopaedics
University of Utah
Salt Lake City, UT, USA

Department of Pathology
University of Utah
Salt Lake City, UT, USA

Department of Bioengineering
University of Utah
Salt Lake City, UT, USA

Department of Physical Medicine and Rehabilitation
Uniformed Services University of the Health Sciences
Bethesda, MD, USA

ISBN 978-3-030-30666-3 ISBN 978-3-030-30667-0 (eBook)
https://doi.org/10.1007/978-3-030-30667-0

This Springer imprint is published by the registered company Springer Nature Switzerland AG
The registered company address is: Gewerbestrasse 11, 6330 Cham, Switzerland

To those who suffer from infection, biofilm-related, or otherwise. Don't lose hope. Solutions are on the horizon as thousands of researchers and clinicians work to resolve infectious problems. Healing will one day transcend pain.
And to my wife, who has never let me quit. She's consistently been there with her hands on my back, when my shoulders have needed squaring and I've had to face the wind.

Foreword

For decades, bacterial and polymicrobial biofilms have been recognized as unique microbial communities, distinct from their planktonic or single sessile microbe relatives. Biofilm phenotypes, behaviors, properties, and persistence survival strategies are now known to be responsible for diverse scientific, medical, and technological problems. Despite substantial published scientific evidence elucidating their diversity and persistence across many environmental niches, microbial biofilms are only now garnering sufficient universal attention and broad scrutiny to categorize them as a unique microbial life-form. Indeed, the biofilm research community continues to advocate specific scientific, pathological, biomedical, and technological differences for established biofilms that are clearly distinct from their individual microbe precursors, either suspended or adherent to surfaces. These differences extend to their physical attributes as living communities, their metabolism, microbiology, pathology, and interactions with materials and within specific ecological, maritime, physiological, and biomedical environments and antimicrobial susceptibility and resistance. Methods to monitor biofilm formation, dynamics, maturation, and behaviors, to control their engagements, and to facilitate their removal and elimination have naturally evolved as biofilm properties are better understood. The resulting implications for controlling biofilms, or eliminating them from specific niches (i.e., as biomedical threats or technological problems), have increasing urgency, given their recognized costly involvement in disease and medical implants as well as their broader technological challenges across their diverse environmental niches.

This book considers the "biofilm problem" from a predominantly medical perspective, with numerous chapters examining current thinking and strategies to diagnose, control, and eradicate pathogenic biofilms from patient wounds and implants. Surgical implantation and clinical placement of medical devices and biomaterials into patients create a predisposition and increased susceptibility to device- and implant-related infections. The result is a catastrophic combination of substantial patient mortality and increasingly unaffordable economic burden. Catheter-related bloodstream and urinary tract infections and periprosthetic joint infections are commonly attributed to biofilm complications. More broadly, orthopedic device-related infections commonly associated with the diversity of implanted biomaterials to repair and augment musculoskeletal tissues represent a daunting clinical challenge. Nonetheless, all implanted materials, whether synthetic (e.g., metallic, polymeric, or ceramic) or natural (e.g., matrix-derived, decellularized, tissue-grafted, chemically fixed, fresh frozen allogeneic, bioreactor-grown), predispose the host (patient) to infections, primarily those of biofilm origin. That clinical biofilms might be prevented as a major source of patient suffering and increasing healthcare cost burden is an important motivation.

Several primary causes of host-implant predispositions to clinical infection are the following: (1) abundance of opportunistic and commensal pathogenic communities and even biofilms within host tissues that are locally disrupted or activated in tissues upon implant placement; (2) recognized pathogenic competence to locate, colonize, and establish biofilm communities on or near biomaterials; and (3) compromised host immunological engagement around an implanted biomaterial, associated with the foreign body response. All implanted medical materials and even many chronic wounds present with these local tissue site compromises, enhancing the risks of patient infection. Incidence and infection risks are empirically observed to be variable, depending on patient health status, comorbidities, biomaterials chemistry and physical attributes (i.e., size, porosity, compliance), tissue physiology or anatomical site of wounding, surgical techniques, and as-yet undetermined predisposing genetic variations, including the commensal microbiome profile, and perhaps opportunistic microbiome features, including endogenous biofilm populations.

Despite increasing recognition that pathogenic biofilm formation is causally linked to many patient infectious disease challenges, both with and without implants (e.g., oral caries and gingivitis, epithelial fungal infections, ocular keratitis, bacterial dermatitis, cystic fibrosis), few clinical antimicrobial therapeutic strategies distinguish virulence factors unique to biofilms from those of related planktonic non-biofilm pathogens and then tailor antimicrobial regimens accordingly. Biofilm formation provides fungal and bacterial pathogens unique physical protection from host immune mechanisms, including frustration of complement activation and phagocyte clearance, and also a barrier against antimicrobial agent exposure. Chemically, biofilm-produced molecular signaling to host microbiome, host tissue cells, and pathogenic partners (e.g., polymicrobial signaling dynamics) effectively confuses host responses, both from commensal protective bacteria (e.g., commensal *Staphylococcus epidermidis*) and endogenous host immune defense mechanisms (e.g., complement lysis, phagocyte uptake, neutrophil-induced free radical activities). Significantly, biofilm-resident bacteria exhibit population phenotypic heterogeneity: some resident colonies are highly pathogenic, while others exhibit reduced metabolic activity (e.g., sleeper cells or persisters) that are inherently refractory to antimicrobials targeting metabolic pathways. This population diversity compromises current clinical anti-infective approaches to mitigate infection. Planktonic pathogens might be readily neutralized by combinations of host clearance and clinical antibiotic regimens, but due to their unique features, biofilm forms do not respond to conventional clinical strategies. Empirically, inhibitory or bactericidal concentrations shown to kill planktonic organisms fail to kill or eradicate biofilms of the same organism. The clinical distinction for addressing mature, colonized, established biofilms as the infection nidus different from conventional "infectious organisms" classically considered as planktonic, invading species must be continuously emphasized. Accordingly, new antimicrobial strategies that address biofilm vulnerabilities must be developed.

The observation that there currently is no formal, universally accepted definition for "biofilm" is troubling. This inability to provide a definition has traditionally caused confusion in the published literature, producing ongoing claims to effectively address biofilm challenges when in fact no biofilm is actually shown. Often, sessile, surface-resident microbes are described as "biofilm" in published work when in fact this simple (and false) characterization fails to consider most of the unique biofilm-specific factors. Recognized traits for biofilms include (1) their unique chemical signaling dynamics (e.g., quorum sensing); (2) their exopolysaccharide barrier matrices comprising both bacterial and host-sourced extracellular products, including nucleic acid nests, that firmly attach these viable colonies to tissue and biomaterials surfaces; (3) their population heterogeneity and persisters; (4) their unique propensity to reside naturally within host tissues without immune recognition; (5) their enhanced physical and biochemical ability to resist both host clearance and antimicrobial assault; and (6) their dangerous capacity to rapidly and readily reseed and repopulate debrided tissues, implants, and necrotic areas with viable satellite biofilm colonies. These specific phenotypic features should define "biofilms" for scientific and research purposes and be further exploited to drive new effective clinical solutions that address biofilm-specific infection vulnerabilities. A clinical "one-size-fits-all" antimicrobial treatment approach to addressing planktonic and biofilm pathogens is clearly losing the battle and might very well be promoting antimicrobial resistance. In this regard, increasing focus on small colony variants that populate biofilms as slow-growing pathogens recognized as major players in difficult, persistent infections is also becoming integral to the ongoing biofilms narrative.

Because reliable identification of implant-centered infection in patients and, more fundamentally, discerning differences between biofilms and planktonic pathogens in clinical infections are compelling unsolved issues, improved biofilm-specific diagnostic tools and methods are required before any rational approach to infection therapy can be decided. Parts of this book also address biofilm diagnostics and infection markers to resolve biofilms as causative and ensure their elimination. Traditional clinical infection markers (e.g., procalcitonin, C-reactive protein, erythrocyte sedimentation rate) are not specific to biofilms. Despite their high sensitivity and routine accessibility, these serum biomarkers currently are used for first-line screening tests for implant infections, particularly in orthopedic implants. However, given patient-specific biomarker variations with infection, effects of patient pharmacologies on these markers, and their intrinsic temporal variability, these markers are often not recommended as sole evidence for implant-associated infections. The lack of sufficiently reliable diagnostic evidence, and ambiguities about specificity and sensitivity of each marker, especially for biofilm infections, motivates the need to develop new approaches. Additionally, the invasive nature of sample procurement from implant sites to confirm infection source, including during infection revision procedures, demands continued focus on discovering and validating reliable surrogate markers or systemic indicators of local, implant-specific infection that are less invasively sampled. Clinical validation of new biomarkers is not trivial: it is time-consuming and expensive. Therefore, diagnosis innovation to improve biofilm analysis and characterization will take time, patience, and research financing to produce progress.

Proper diagnosis and molecular profiling of biofilms, both in laboratory preclinical work and in patients, will improve the reliability of information and insights necessary to facilitate development of more effective strategies to find, study, probe, and control biofilms. Fundamental discoveries regarding biofilm microbiology and phenotypes will guide better methods to modulate and eradicate them. Pairing detection and analysis to identify biofilm-specific infections will better guide therapeutic management of biofilm-centered infection, a sorely needed improvement to benefit many patient infections, and especially those involving biomaterials and implants.

David W. Grainger
Departments of Biomedical Engineering
and of Pharmaceutics and Pharmaceutical Chemistry
University of Utah,
Salt Lake City, UT, USA

Preface

This isn't an autobiography, but I hope you'll excuse an autobiographical note as I try to set the stage for this book and provide a backdrop of my experience, as it provides a basis for why I propose a targeted approach against the biofilm phenotype (where biofilm is applicable) in translational research, antimicrobial testing, and device development.

My introduction to biofilms came as an undergraduate student. During a Microbial Ecology course, my professor spent a few brief moments during a lecture to discuss observations of bacteria living in clusters, but little was discussed about the impact of biofilms on the environment or its role in ecology. It surprises me now that the predominant dwelling state of bacteria in natural ecosystems was touched on only briefly in a microbiology program, but despite the brevity, I was instantly intrigued. At the time, I didn't realize biofilm-related research would become the foundation of my career, but I'm glad it did.

Three years after I graduated with a bachelor's degree in Microbiology, I was hired as a member of the first team in the United States developing osseointegrated (OI) implant technology for wounded warriors. In brief, OI implants provide an alternative to socket prosthetic technology. They consist of a metal post that is surgically inserted into the residual bone (e.g., femur) of a patient with a portion that protrudes through the skin for prosthetic limb attachment. This design provides direct load-bearing to the skeleton as opposed to soft tissues and reduces bulkiness. However, the percutaneous nature and solid surfaces of OI implants make them innately challenged as they are particularly well-suited to host and harbor biofilms, which thrive in air/solid/liquid interfaces. I was tasked with supporting the development of antimicrobial strategies to treat and prevent biofilm implant-related infections at OI implant interfaces. Several years of data collection showed that a mechanical skin seal established by the host may be the principal method of controlling infection outcomes. Data further indicated that a daily hygienic wash, similar to brushing teeth, may serve as a secondary method to help manage biofilm burden and reduce the risk of infection.

I shifted my efforts as a graduate student to the development of an active release antimicrobial coating that could be applied to orthopedic implants, with an emphasis on fracture fixation plates. As I prepared the study design, I considered what I had learned about biofilms, their predominance in natural ecosystems and impact on device-related infection. I wondered, "If estimates suggest that >99% of bacteria in natural environments, such as dirt and soil, dwell in the biofilm phenotype, wouldn't that mean that when an injured soldier or a civilian patient breaks open their leg, the wound would be contaminated with biofilms at the point of injury, and not just planktonic bacteria?" It was apparent after scouring the literature that essentially all animal models of infection involved inoculation with planktonic bacteria. But that didn't seem to me to model an open fracture situation, wherein mature bacterial biofilms from dirt, mud, or soil would be the initial contaminants. I proposed the idea of developing an active release antimicrobial coating that specifically targeted the biofilm phenotype in an animal model of open fracture infection to my mentor, Roy Bloebaum, PhD, and two other colleagues, Peter Beck, MD, and John Hibbs, MD. Within a year, we had secured NIH funding to test a unique coating in a sheep model of open fracture infection—using biofilms as initial inocula. In short, data indicated that established biofilms were able to cause low-lying infection that was chronic in nature and positive infection control animals were able to survive to the endpoint without antibiotic intervention. Furthermore, our antibiofilm coating, which we first optimized against biofilms in vitro, was able to prevent biofilm implant-related infection in 100% of cases.

It was an honor to host William (Bill) Costerton, PhD (1934–2012), in our lab for 2 days toward the end of my graduate school experience. Bill (he didn't like me calling him Dr. Costerton) is considered by many to be the father of biofilm discovery and research. I had shared with him the concept of using biofilms as initial inocula in animal models of infection, and we published a paper on the topic. Bill Costerton and my wife were the two catalysts who gave me the confidence to work toward the completion of my PhD. Bill was a champion of students. My graduate experience had been particularly challenging, and I was on the verge of quitting. Bill and my wife talked me out of it. It was particularly humbling when he told me that my graduate project was one of the top ten graduate projects he had encountered in his career (he probably said that to every graduate student). His validation gave me an extra sense of commitment. I worked harder than ever after our visit; I memorized the contents of more than 300 scientific papers, so I could defend my dissertation, completed my 10th and 11th papers, authored a book chapter, and practiced my presentation skills on a weekly basis until I could stand with confidence in front of my committee and show them I was worthy of the PhD. In the end, it worked out. The reason I share this isn't to inflate myself. Rather, I hope my experience resonates with graduate students, in particular those who may be struggling—the effort is worth it!

The data collected during my graduate years was promising, and we had hopes of advancing our unique antibiofilm coating to the clinic. However, within months of receiving my PhD, the technology transfer office of my university sat me down and told me that I could no longer work on the technology I had spent 5 years developing. Legal documents had not been established correctly between several parties. That was a challenging time for me as a newly minted investigator and forced me to pivot my biofilm research efforts. But I soon realized that receiving a PhD wasn't the endgame, it was just the beginning. There are myriad amounts of studies that need to be done to address the problem of biofilm-related infection. And this research can't be accomplished in one study, by just one lab or by a single organization.

As evidenced in this book, many approaches are being taken to tackle the biofilm problem, and our collective efforts are consistently increasing the potential for technologies and protocols to manage biofilm-related infection. I never considered that entering into the biofilm space would be so rewarding, but it has led to the accomplishment of my aspiration of being a professor at the University of Utah and more: directing a robust lab in the Department of Veterans Affairs with enthusiastic students and postdoctoral fellows, all of whom are engaged with me and thousands of others in research institutes and companies across the globe in the pursuit of antimicrobial technologies that can target the biofilm phenotype. Our most important work may be on the horizon wherein lies the culmination of therapies put into clinical practice where they can improve quality of life and relieve suffering. And perhaps, the time will come when I can share a copy of this book with my undergraduate microbial ecology professor. It will certainly provide many discussion points that can be considered for far longer than a few moments of a lecture.

Salt Lake City, UT, USA Dustin L. Williams

Acknowledgments

Thank you to those who helped me start my career and who took a chance on me early on. I recognize that the efforts of many have served as pivot points in my career, ultimately making this book possible. In a random encounter that lasted only a few minutes, Doug Powell, MD, was kind to listen to me as a young 20 something when I expressed an interest in becoming a scientist. Within a day, he put me in touch with his colleague, Sancy Leachman, MD, who allowed me to volunteer in her lab and showed me the ropes of research. She helped me find my first paid position with Ray Warters, PhD. His guidance provided me with a skill set that led to a research position with Roy Bloebaum, PhD, and team members including Cathy Petti, MD. Roy was driven to develop the first transfemoral osseointegrated implant in the United States. He and many team members had their hands full with an aspiring scientist who, admittedly, could be (over)ambitious at times. We often joked in the lab that Roy only gambles on black 17 (in roulette), but he metaphorically stepped outside his comfort zone and took a chance on me as a PhD student. I'm grateful he did and am indebted to him and Charles Saltzman, MD, for the opportunity to take over the Bone and Joint Research Lab after completing my PhD training. A heartfelt thank you also goes to the many students, technicians, project supervisors, postdoctoral fellows, volunteers, and administrators who have put in late nights, weekends, holidays, and everything in between to make our work possible.

Special thanks go to Brad Isaacson, PhD, MBA, MSF, PMP, who grunted with me through Graduate School and has become one of my dearest friends, collaborators, and colleagues. We have both come a long way from eating sandwiches at Subway to now watching our kids bounce around toys at local playgrounds or working late into the night to make sure grant applications are polished and poised for submission.

Thank you to many in the fields of biofilm research. One of my first encounters with a biofilm researcher was with Al Parker, PhD, in Cancun, Mexico. He indulged me in conversation as I sipped on a can of soda at the Hilton Cancun Beach and Golf Resort during a triannual ASM Conference on Biofilms. He and Darla Goeres, PhD, offered insights into the development of a modified biofilm reactor that would allow me to use biofilms as initial inocula in a unique animal model. Bill Costerton, PhD, was likewise gracious and willing to support my ideas and even visited our lab to help guide me through several aspects of my dissertation. I'll never forget his parting words, which can't be printed in a professional book such as this but are unforgettable. I thank Garth James, PhD, for years of published and verbal insights into methods that underlie medically relevant biofilm experiments. I, and several of my lab members, also thank Phil Stewart, PhD, for making an effort to bring the world of biofilms to life with videos that capture an audience and broaden our imagination of what can be accomplished in biofilm research.

Thank you to Javad Parvizi, MD, for engaging me and my department members in multiple biofilm-relevant conversations and for offering me my first opportunity to contribute a chapter in a book. Thank you to Peter Beck, MD, and John Hibbs, MD, who spent hours listening to my experimental design considerations, data dumps, and presentations. Thank you to Richard Tyler Epperson and Brooke Kawaguchi for being the backbone of our lab. Thank you to Ryan Davies; Todd Kinard, JD; Ryan Looper, PhD; Paul Sebahar, PhD; Travis Haussener, PhD; and Chad Testa, PhD, for synergizing efforts to tackle the biofilm problem with new antibiofilm compound development. Similarly, thank you to Jeff Sivertsen, JD; Nicholas Ashton, PhD; Jason Rivkowich; Christophe Courcimault, PhD; Shannon Kegel; and Michaela Rivkowich for their willingness and time to develop a unique device that targets biofilms. There are too many others to list by name, but a heartfelt thank you to the many researchers, collaborators, and colleagues who take the time to communicate by email or visit at conferences, meetings, and other gatherings.

Thank you also to Merry Stuber and the team at Springer Nature for reaching out to me for a book proposal. Finally, thank you very much to the fantastic coauthors of this book who worked nights, weekends, holidays, and more to pull the chapters together. The insights provided by this collective group of authors shed light on the biofilm problem from multiple vantage points including private and public research, medical practice, military medicine and healthcare, industry, and academia. You have achieved my vision for this project. Your efforts are appreciated, noticed, and will make this text an important contribution to the field.

Contents

We Begin to Target the Biofilm

Dustin L. Williams

Abstract Biofilms were unknowingly observed centuries ago. Within the last 70 years, our understanding of biofilms, their morphology, and characteristics has grown. Yet despite an increased understanding, the presence of biofilms and their impact on healthcare are still often overlooked, or misunderstood. Antimicrobial technologies and applications are primarily focused on planktonic bacteria. A targeted approach against the biofilm phenotype is likely to improve infection outcomes wherein biofilms are the source of difficult-to-treat infections.

Keywords Biofilm history · Unexploited opportunity · Target · Biofilm phenotype · Initial inocula · Antibiofilm

A Brief Timeline of Biofilm Discovery and Characteristics

Our knowledge of biofilms has grown exponentially over the past several decades. Of course, almost obligatorily any biofilm conversation has to involve mention of the initial observations made by Antonie van Leeuwenhoek nearly 400 years ago as he viewed scrapings of his teeth through hand-made microscopes. His drawings of circular and rod-shaped microorganisms are now commonplace in textbooks, but his findings at the time were an unrecognized hinge upon which the gates of

D. L. Williams (✉)
Department of Veterans Affairs, Salt Lake City, UT, USA

Department of Orthopaedics, University of Utah, Salt Lake City, UT, USA

Department of Pathology, University of Utah, Salt Lake City, UT, USA

Department of Bioengineering, University of Utah, Salt Lake City, UT, USA

Department of Physical Medicine and Rehabilitation, Uniformed Services University of the Health Sciences, Bethesda, MD, USA
e-mail: dustin.williams@utah.edu

D. L. Williams (ed.), *Targeting Biofilms in Translational Research, Device Development, and Industrial Sectors*,
https://doi.org/10.1007/978-3-030-30667-0_1

microbiological discovery would swing. Van Leeuwenhoek's unearthing of microscopic life—which had been proposed for nearly a century prior by others including Girolamo Fracastoro—didn't appeal to the masses of his time or receive general credence given the ensuing debate of Aristotelean abiogenesis (i.e., spontaneous generation). Yet with hindsight as our luxury, we now link his observations with those made more recently and consider his visual findings to be the first indicator of a complex microscopic life system—now known as biofilms.

More recent milestones in biofilm discovery include Claude ZoBell's sea water work, which indicated that bacteria preferentially adhere to solid surfaces [1]. Thirty-five years after ZoBell's publication, J. William (Bill) Costerton and colleagues published a seminal paper describing how bacteria stick tenaciously to said solid surfaces and produce microcolonies of sister cells via glycocalyx components that facilitate intercellular and extracellular adhesion [2]. Throughout the 1970s, physicians noted bacterial aggregates in the sputum of patients who suffered from cystic fibrosis [3]. The link between infection etiology and the communal nature of bacteria in natural ecosystems began to find correlation. The term "biofilm" was formalized between 1978 and 1981 and is used to define a dynamic community of bacteria—more accurately "a microbially derived sessile community characterized by cells that are irreversibly attached to a substratum or interface or to each other, embedded in a matrix of extracellular polymeric substances that they have produced, and exhibit an altered phenotype with respect to growth rate and gene transcription [4]." This phenotype contrasts the planktonic, or individually free-floating bacterial phenotype, such as may be found in a laboratory test tube.

In the early to mid-1980s, biofilms were found to be complicating agents in medical device-related infections. More specifically, biofilms seemed to serve as reservoirs of infection: although systemic antibiotic therapies quelled acute symptoms, infection would recur once antibiotics were discontinued [5, 6]. This suggested that antibiotic regimens eradicated planktonic bacteria that dispersed away from the biofilm reservoir, but they failed to eradicate the source. The chronic and difficult-to-treat nature of these pathologies resolved once the nidus of infection was removed (in these cases the nidus was an implanted device on which biofilms had formed). The recalcitrant nature of biofilm-related infection began to be clarified in 1985 as bacteria in biofilms were found to have inordinate antibiotic tolerance profiles compared to planktonic counterparts [7]. Many considered that biofilms were more tolerant to antibiotics because of extracellular matrix components "gumming up" the diffusive capacity of antibiotics, preventing them from reaching the tortuous interstices of the biofilm core. This presumption was short-lived as by 1988, small molecules were shown to penetrate to the core of a biofilm in a matter of seconds [8].

Elucidations of the eclectic nature of biofilms didn't stop there. By 1991 water channels were observed in biofilms—facilitating diffusion of nutrients, waste products, and signaling factors—further revealing the intricacies of their dynamic three-dimensional structures [9]. And in the early 2000s, oxygen gradients were found to develop in biofilms, resulting in a profile of aerobic cells toward the outer regions and anaerobic cells in the core [10]. Perhaps conspicuously the most common biofilm-forming pathogens are facultative anaerobes. If all of the other characteristics of biofilms weren't enough, the facultative anaerobic characteristic of biofilms makes them particularly menacing. As aerobically dividing cells predominate in the outer regions of the community, they can "gobble up" infiltrating attacks (such as antibiotics) as well as utilize free oxygen. This results in anaerobically dividing cells predominating in the biofilm core. In an anaerobic state, cells in the core are less susceptible to antibiotics, as antibiotics typically function against metabolically active and dividing cells. The core-dwellers also require less nutrients as they function in a more sessile state. This dual aerobic-anaerobic nature of living allows the biofilm to switch back and forth between growth states, adjusting to the environment "on the fly." It also seems to suggest that the aerobically dividing cells in the exterior are always ready (at least from my perspective) to protect recalcitrant variant cells in the inner core from external perturbations, much like a bee swarm protects its queen.

Unexploited Opportunity to Target Biofilms in Healthcare

The tenacity of biofilms to survive and thrive in every domain on Earth reflects the evolutionary advantage that bacteria have by clustering and congregating together. Most of what we see published about biofilms today builds upon the early discoveries discussed. Yet despite the discoveries of the past 70 years, and the increasing acceptance of the biofilm theory, a recognition of biofilms in clinical care is still not commonplace. As detailed in the final chapter of this book, all antibiotics that are developed, tested, and optimized, all antibiotic dosing regimens that are established, all clinical reference laboratory standards, and all medically relevant regulatory guidelines are based solely on outcomes related to planktonic bacteria. Further, there is not yet a type of indwelling medical device that is not affected by biofilm formation and accompanying infection in at least a subset of instances, yet biomedical engineers and companies rarely take infection into account during device development. Granted, the current regulatory environment is not conducive to combination products or antibiofilm technologies (discussed by Armbruster in chapter "Biofilm Infections in Orthopedic Surgery and Their Impact on Commercial Product Development"). Nevertheless, with the exception of a few companies (such as Next Science in chapter "Targeting Biofilms in Orthopedic Infection") and a growing number of academic and clinical pursuits (see chapters "Catheter Associated Urinary Tract Infections: Development of a Test Method for Assessing the Efficacy of Antimicrobial Technologies/Products" and "Biofilms and Wound Infection Research in the U.S. Military"), as a society we have not yet begun to target the biofilm as part of our antimicrobial strategy regimens. Is the failure to target biofilms affecting clinical outcomes? In light of the growing body of evidence over the last 70 years, we can answer with a resounding, "Yes!"

The chapters in this book collectively provide multiple examples wherein targeting biofilms could potentially improve biofilm-related infection outcomes. As a primer, let's consider a specific example: the paradigm of human skin and surgical prep solutions. Human skin harbors polymicrobial biofilms that dwell deep, even beneath the subepidermal compartments [11, 12]. A common example includes *Cutibacterium acnes* (formerly *Propionibacterium acnes*), which forms mature biofilms in sebaceous glands and hair follicles [11]. This particular organism has been increasingly problematic in particular with shoulder surgeries [13–15], shoulder arthroplasty [16, 17], and spine surgeries [18–20]. We'll revisit this information about biofilms dwelling deep in human skin after a discussion on surgical skin prep solutions.

Across the globe there are hundreds of thousands of surgeries performed each day that require a preoperative skin prep. Joseph Lister used carbolic acid in the first skin prep solution circa 1867 [21, 22]. Today, prep solutions primarily contain povidone-iodine (PI) or chlorhexidine gluconate (CHG) in combination with isopropyl alcohol. PI is a relatively poor disinfectant with minimum inhibitory concentration (MIC) values over 512 μg/mL against staphylococci and many other species but is relatively nontoxic when used topically and is used at high concentrations (7.5% (75,000 μg/mL)–10% w/v (100,000 μg/mL)) that allow it to achieve acceptable disinfectant outcomes. CHG is more potent, with MIC values in the single-digit range (e.g., 1–2 μg/mL) against a broad spectrum of organisms, but is more toxic and thus is typically less concentrated at 4% w/v (40,000 μg/mL). These disinfectants are typically used in combination with isopropyl alcohol ranging from 4% to 74%.

When used as preoperative surgical preps, solutions are slathered on the skin in a distinct inward-outward pattern to sterilize the surgical site of interest. The application procedure and dwell time of each solution is highly variable, depending on logistics, environment, and training regimen of personnel in each operating room/organization. Yet general directions for use are provided. For example, directions for using HIBICLENS® (4% CHG + 4% isopropyl alcohol) in preoperative skin preparation are:

> Apply HIBICLENS liberally to surgical site and swab for at least 2 minutes. Dry with a sterile towel. Repeat. Dry with a sterile towel.

In the case of Betadine® Surgical Scrub (7.5% povidone-iodine), directions are:

> Wet skin with water. Apply Scrub (1 cc is sufficient to cover an area of 20–30 square inches); develop lather and scrub thoroughly for about 5 minutes. Rinse off using sterile gauze saturated with water. The area may then be painted with Betadine solution and allowed to dry.

The protocol for HIBICLENS provides a dwell time of approximately 4 minutes, if the directions are followed. Betadine provides a dwell time (including dry time) of approximately 10 minutes. There is evidence that Betadine and CHG can bind proteins and remain on skin for 4–24 hours, respectively, but we should consider that the disinfectant would be in a diluted and nonaqueous state if the solution is rinsed or dried off. Product applications are also limited to the top layers of skin.

There are no data showing that CHG is effective against skin-dwelling biofilms within 4 minutes. Time kill studies have been performed but against planktonic cells [23]. As mentioned, all antimicrobial susceptibility profiles that are established and regulated are based on planktonic bacterial measures (e.g., the MIC). Furthermore, data that are collected are typically done so in static broth conditions, not in complex skin environments, with multihour endpoints, not minutes. There are some data showing extended dwell times (up to 24 hours) with PI (again in static broth conditions) [24] but focus on the inhibitory effect of PI on biofilm formation, not eradication of well-established biofilms—the phenotype in which bacteria dwell in skin.

With this information we can begin to make a connection between a common clinical procedure—preoperative surgical prep—and the potential impact that the lack of a targeted approach against the biofilm phenotype might have. Surgical prep solutions may be effective at eradicating planktonic bacteria that dwell in the upper regions of skin, but data is lacking on their effect on well-established biofilms in these same regions. Furthermore, preoperative skin prep solutions are specific to the top layers of the skin, such as the stratum corneum, and may not penetrate rapidly into the deeper layers where established biofilms dwell, such as those formed by *C. acnes* [11, 25]. Thus, as a surgeon makes a cut with a scalpel during surgery, subepidermal or deep-layer biofilms that are unaffected by preoperative skin prep could be mechanically driven from the host's deep skin layers to the tissues of the surgical site.

As we observe an increase in the number of *C. acnes*-related infections in shoulder surgeries (cited above), pacemaker implantation [26], and spine surgeries (cited above), and as we continue to see staphylococcal species predominate in implant-related infections [27–29], we should consider designing experiments that assess the influence of biofilms in the patient's own skin that could contaminate a surgical site during a procedure. To further the argument, normal flora organisms have been shown to colonize wound borders within 30–180 minutes following surgical prep [30, 31]. This supports what is well-known among surgeons: the longer a surgical procedure takes, the higher the risk of bacterial contamination and potential infection.

My lab team recently wanted to collect in-house data to test the hypothesis that deep-dwelling, potentially subepidermal bacteria in skin regions can survive a preoperative surgical prep. We prepped the skin on two pigs during an excision wound surgery (IACUC approved) using alternating treatments of Betadine (10%) and isopropyl alcohol (70%). We sterilely draped the pig and prepped the skin one last time with Chloraprep™. After ~10 minutes of dry time we biopsied sixteen skin samples (eight from each pig) that reached the pig fascia, placed them sterilely into broth, ground the tissues and cultured samples under aerobic and anaerobic conditions. Data in this limited pilot set showed that our hypothesis was supported; more than five isolates were identified with ~10^5–10^6 CFU/g tissue, suggesting that a thorough skin prep procedure may not affect bacteria that dwell deep in host skin (publication pending).

Strategies are being employed to account for the limitations of preoperative surgical preps. Surgeons are having patients wash their surgical site with either CHG or benzoyl peroxide-based products the night before surgery. This is consistent with recommendations from the Centers for Disease Control (CDC), which suggests patients can shower or bathe in chlorhexidine the night before surgery. The idea behind these practices is to increase dwell time and surface area coverage of an antimicrobial on the skin to reduce the risk of normal flora organisms—primarily dwelling in biofilms—from causing opportunistic infection. Another strategy (from personal communications with surgeons) is the practice of using a "dirty" and "clean" scalpel during a procedure. Surgeons make an initial cut in the skin, and recognizing that the first scalpel may be contaminated with deep-dwelling bacteria, they toss it aside and use one or multiple clean scalpels to complete the procedure.

A needle-in-a-needle device has been patented to achieve a similar goal in eye surgeries [32]. This device has an outer, hollow bore needle in which a secondary needle is slightly retracted. As the first needle punctures the site of interest it interacts with bacteria (including those in biofilms) that may be present in the outer regions of host tissue. The second needle is progressed through the hollow bore of the first needle and can extend into deeper tissue regions with reduced risk of introducing bacterial contaminants that are "stuck" on the outer cylinder of the first needle. This dual-needle device concept is interesting as it could be applied to other biofilm-relevant paradigms.

Skin is only one of many anatomical sites colonized with established biofilms. The nares, colon, sinus tract, and oral cavity are all colonized by a complex microbiome that primarily consists of biofilm-dwelling organisms. Exogenous materials including soil, particulates, animals, and foreign bodies can also harbor biofilms that can interact with us in a variety of circumstances (see final chapter of this book). As we develop medical devices, antimicrobial technologies, or therapeutics, each domain can be considered in the context of biofilms to determine how we can exploit and target this phenotype of bacteria and improve patient outcomes [33].

Not to Forget Biofilm Inhibition

This chapter and book as a whole focus primarily on the effects of biofilms on infection outcomes. This isn't to suggest that we shouldn't consider the importance of inhibiting biofilm formation, which is an additional approach toward preventing biofilm-related infection. However, I've chosen to focus on well-established biofilms primarily based on conversations with surgeons, healthcare workers, and researchers. The sequential, staged model of biofilm formation (stages of biofilm formation are presented by Haymond in chapter "Translation of Antibiofilm Technologies to Wounds and Other Clinical Care") is often strongly adhered to; many have been taught and purport that in order for a biofilm to be present, planktonic bacteria must first adhere to a surface, then form a biofilm, and thus planktonic bacteria are really the only phenotype to consider when developing an infection treatment strategy. As a specific example, I was presenting data at a meeting from one of our sheep models wherein we first grow biofilms then inoculate them in the animal at the time of surgery [34, 35]. An audience member asked at what point we introduce the planktonic bacteria that will form a biofilm and cause infection? In our model, the biofilm is already there, just as biofilms are already present in natural ecosystems [36], and established biofilms can cause infection [34, 35]. We may need an expansion in our thinking to consider there are times when a sequential process of biofilm formation isn't relevant if bacteria in an environment are already in an established biofilm (see discussion on open fractures and contamination with biofilm-containing soil in the final chapter). Such is the case presented above of preoperative skin preps and established biofilms in the epidermal and dermal layers.

When Is a Biofilm

I was attending the 8th ASM Conference on Biofilms in Washington, DC, in the fall of 2018, and watched videos of neutrophils attempting to gobble up small clusters of bacterial cells. The purpose of the videos was to show that within 4 hours of growth, bacterial aggregates had developed to a point that neutrophils could not phagocytose them (whether it was the mass of the aggregates or slime production is not yet known). As I watched the videos, I had this thought, "We focus heavily on the characteristics that define a biofilm. Yet it may be important to consider not just what a biofilm is, but when a biofilm is." The scientific and clinical community considers, as a rule of thumb, that a biofilm is mature by 24 hours. This has traditionally been based on a time point when bacteria in a biofilm begin to display increased tolerance to antibiotics and display signs of maturity such as matrix production and/or three-dimensional structure formation. Yet if small bacterial aggregates can evade the swallows of neutrophils so early (granted, this work was on solid slides and not in the body), perhaps we consider that the maturity of a biofilm is based on more than its tolerance factors to antibiotics, but also its ability to evade host immunity. For years, Dave Armbruster (see chapter "Biofilm Infections in Orthopedic Surgery and Their Impact on Commercial Product Development") and I have tossed around the question of what impact low-number biofilms might have on infection outcomes. In-house data from my lab has shown that low-number biofilms with a mere 10^3 CFU can be 1000× more tolerant to antibiotic treatment compared to planktonic counterparts (unpublished data). Could low-number biofilms show similar recalcitrance to host immunity? And are low-number biofilms more relevant in clinical paradigms than the gregarious amounts of bacteria we typically inoculate in animal models? We hope to find out.

On the other side of the bookend, data being collected in our lab indicate that some isolates of methicillin-resistant *S. aureus* (MRSA) may not begin to form measurable biofilms until ~192 hours of growth in a reactor system that has been optimized for biofilm growth, whereas others flourish (develop plume heights of >50 μm) in ~96 hours [37]. Thus, the "when" a biofilm forms won't be the same for every organism. It is likely genus, species, or even strain-dependent, and the growth environment and substrate also play pivotal roles. Overall, as we continue to uncover additional characteristics of biofilms (or bacterial aggregates)—such as optimizing timing of antimicrobial strategies—we'll open the door to new opportunities for chemical, physical, electromagnetic, or other methods that could improve infection control measures.

Teach an Old Antibiotic New Tricks

Although doors may lead to new opportunities, we may also need to reopen old doors that contain uncovered information. Specifically, another consideration for targeting the biofilm phenotype is to assess old antibiotics or other compounds in new ways [38]. All clinical reference laboratories and Clinical and Laboratory Standards Institute (CLSI) protocols define antibiotic efficacy based on screening against planktonic bacteria, primarily using the minimum inhibitory concentration (MIC) value. The limitations of basing antibiotic efficacy solely on MIC values will be discussed in detail in the final chapter, but we'll strike the reader's curiosity here by postulating that perhaps there are antibiotics, or other compounds for that matter, that have been screened with miserably poor MIC values, and have thus been tossed aside as ineffective agents, but that might be more effective against bacteria in biofilms than those in the planktonic state. Our lab has made a recent discovery to suggest just that; specific types of presumably ineffective antibiotics (by MIC standards) are more effective against bacteria in the biofilm phenotype than those in the planktonic state. This reopens a door to consider what improvements could be made with available antibiotics in biofilm-related infection treatment if screening assays were expanded to include efficacy profiles against biofilm-dwelling organisms, in addition to planktonic bacteria.

We Are Gaining Ground but There Is Work to Do

Clinicians, researchers, companies (industrial and healthcare), and technicians have never understood the biofilm problem better than they do today. We understand the mechanics, impact, and weaknesses of biofilms with more detail than ever before. But there is still work to do. We're only at the precipice of understanding the complexities of biofilms, and we face an even greater challenge: there is still a paucity of antimicrobial technologies and protocols that target the biofilm phenotype. As we continue to make advancements in areas of discovery and innovation, we need to pave a parallel path of informing those on the front lines (e.g., nurses, surgeons, podiatrists) about the biofilm theory and incorporate antibiofilm strategies into daily healthcare regimens, with an overall goal of reducing morbidity and mortality that far too often are the result of biofilm-related infection. My hope is that this book adds to the many efforts that are under way to make that happen.

Disclaimer The opinions and information presented in this chapter are those of the author and do not necessarily reflect the position or policy of the Department of Veterans Affairs or the US Government.

References

1. ZoBell, C. E. (1943). The effect of solid surfaces upon bacterial activity. *Journal of Bacteriology, 46*, 39–56.
2. Costerton, J. W., Geesey, G. G., & Cheng, K. J. (1978). How bacteria stick. *Scientific American, 238*, 86–95.
3. Hoiby, N. (2017). A short history of microbial biofilms and biofilm infections. *APMIS: acta pathologica, microbiologica, et immunologica Scandinavica, 125*, 272–275. https://doi.org/10.1111/apm.12686.
4. Donlan, R. M., & Costerton, J. W. (2002). Biofilms: Survival mechanisms of clinically relevant microorganisms. *Clinical Microbiology Reviews, 15*, 167–193.
5. Marrie, T., Nelligan, J., & Costerton, J. (1982). A scanning and transmission electron microscopic study of an infected endocardial pacemaker lead. *Circulation, 66*, 1339–1341.
6. Gristina, A. G., & Costerton, J. W. (1984). Bacteria-laden biofilms: A hazard to orthopedic prostheses. *Infections in Surgery, 3*, 655–662.
7. Nickel, J. C., Ruseska, I., Wright, J. B., & Costerton, J. W. (1985). Tobramycin resistance of *Pseudomonas aeruginosa* cells growing as a biofilm on urinary catheter material. *Antimicrobial Agents and Chemotherapy, 27*, 619–624.
8. Nichols, W. W., Dorrington, S. M., Slack, M. P. E., & Walmsley, H. L. (1988). Inhibition of tobramycin diffusion by binding to alginate. *Antimicrobial Agents and Chemotherapy, 32*, 518–523.
9. Lawrence, J. R., Korber, D. R., Hoyle, B. D., Costerton, J. W., & Caldwell, D. E. (1991). Optical sectioning of microbial biofilms. *Journal of Bacteriology, 173*, 6558–6567.
10. Borriello, G., et al. (2004). Oxygen limitation contributes to antibiotic tolerance of *Pseudomonas aeruginosa* in biofilms. *Antimicrobial Agents and Chemotherapy, 48*, 2659–2664.
11. Brandwein, M., Steinberg, D., & Meshner, S. (2016). Microbial biofilms and the human skin microbiome. *NPJ Biofilms and Microbiomes, 2*, 3. https://doi.org/10.1038/s41522-016-0004-z.
12. Grice, E. A., et al. (2008). A diversity profile of the human skin microbiota. *Genome Research, 18*, 1043–1050. https://doi.org/10.1101/gr.075549.107.
13. Clark, J. J. C., Abildgaard, J. T., Backes, J., & Hawkins, R. J. (2018). Preventing infection in shoulder surgery. *Journal of Shoulder and Elbow Surgery, 27*, 1333–1341. https://doi.org/10.1016/j.jse.2017.12.028.
14. Atesok, K., et al. (2017). Postoperative deep shoulder infections following rotator cuff repair. *World Journal of Orthopedics, 8*, 612–618. https://doi.org/10.5312/wjo.v8.i8.612.
15. Kadler, B. K., Mehta, S. S., & Funk, L. (2015). Propionibacterium acnes infection after shoulder surgery. *International Journal of Shoulder Surgery, 9*, 139–144. https://doi.org/10.4103/0973-6042.167957.
16. Fink, B., & Sevelda, F. (2017). Periprosthetic joint infection of shoulder arthroplasties: Diagnostic and treatment options. *BioMed Research International, 2017*, 4582756. https://doi.org/10.1155/2017/4582756.
17. Nelson, G. N., Davis, D. E., & Namdari, S. (2016). Outcomes in the treatment of periprosthetic joint infection after shoulder arthroplasty: A systematic review. *Journal of Shoulder and Elbow Surgery, 25*, 1337–1345. https://doi.org/10.1016/j.jse.2015.11.064.
18. Georgy, M., Stern, M., & Murphy, K. (2017). What is the role of the bacterium Propionibacterium acnes in type 1 Modic changes? A review of the literature. *Canadian Association of Radiologists journal = Journal l'Association canadienne des radiologistes, 68*, 419–424. https://doi.org/10.1016/j.carj.2017.07.004.

19. Ganko, R., Rao, P. J., Phan, K., & Mobbs, R. J. (2015). Can bacterial infection by low virulent organisms be a plausible cause for symptomatic disc degeneration? A systematic review. *Spine, 40*, E587–E592. https://doi.org/10.1097/BRS.0000000000000832.
20. Uckay, I., et al. (2010). Spondylodiscitis due to Propionibacterium acnes: Report of twenty-nine cases and a review of the literature. *Clinical Microbiology and Infection, 16*, 353–358. https://doi.org/10.1111/j.1469-0691.2009.02801.x.
21. Lister, J. (1867). On a new method of treating compound fracture, Abscess, &c., with observations on the conditions of suppuration. *Lancet*, 336–339.
22. Worboys, M. J. (2013). Lister and the performance of antiseptic surgery. *Notes and Records of the Royal Society of London, 67*, 199–209. https://doi.org/10.1098/rsnr.2013.0028.
23. Edmiston, C. E., Jr., et al. (2013). Reducing the risk of surgical site infections: Does chlorhexidine gluconate provide a risk reduction benefit? *American Journal of Infection Control, 41*, S49–S55. https://doi.org/10.1016/j.ajic.2012.10.030.
24. Oduwole, K. O., et al. (2010). Anti-biofilm activity of sub-inhibitory povidone-iodine concentrations against Staphylococcus epidermidis and Staphylococcus aureus. *Journal of Orthopaedic Research, 28*, 1252–1256. https://doi.org/10.1002/jor.21110.
25. Karpanen, T. J., et al. (2008). Penetration of chlorhexidine into human skin. *Antimicrobial Agents and Chemotherapy, 52*, 3633–3636. https://doi.org/10.1128/AAC.00637-08.
26. Okuda, K. I., et al. (2018). The composition and structure of biofilms developed by Propionibacterium acnes isolated from cardiac pacemaker devices. *Frontiers in Microbiology, 9*, 182. https://doi.org/10.3389/fmicb.2018.00182.
27. Oliveira, W. F., et al. (2018). Staphylococcus aureus and Staphylococcus epidermidis infections on implants. *The Journal of Hospital Infection, 98*, 111–117. https://doi.org/10.1016/j.jhin.2017.11.008.
28. Montanaro, L., et al. (2011). Scenery of Staphylococcus implant infections in orthopedics. *Future Microbiology, 6*, 1329–1349. https://doi.org/10.2217/fmb.11.117.
29. Tong, S. Y., Davis, J. S., Eichenberger, E., Holland, T. L., & Fowler, V. G., Jr. (2015). Staphylococcus aureus infections: Epidemiology, pathophysiology, clinical manifestations, and management. *Clinical Microbiology Reviews, 28*, 603–661. https://doi.org/10.1128/CMR.00134-14.
30. Achermann, Y., Goldstein, E. J., Coenye, T., & Shirtliff, M. E. (2014). Propionibacterium acnes: From commensal to opportunistic biofilm-associated implant pathogen. *Clinical Microbiology Reviews, 27*, 419–440. https://doi.org/10.1128/CMR.00092-13.
31. Johnston, D. H., Fairclough, J. A., Brown, E. M., & Morris, R. (1987). Rate of bacterial recolonization of the skin after preparation: Four methods compared. *The British Journal of Surgery, 74*, 64.
32. Olson, R. J. (2012). Hypodermic needle system and method of use to reduce infection. USA patent.
33. Williams, D. L., & Costerton, J. W. (2011). Using biofilms as initial inocula in animal models of biofilm-related infections. *Journal of Biomedical Materials Research, 100*, 1163–1169.
34. Williams, D. L., et al. (2012). *In vivo* efficacy of a silicone – cationic steroid antimicrobial coating to prevent implant-related infection. *Biomaterials, 33*, 8641–8656.
35. Williams, D. L., et al. (2012). Experimental model of biofilm implant-related osteomyelitis to test combination biomaterials using biofilms as initial inocula. *Journal of Biomedical Materials Research Part A, 100*, 1888–1900.
36. Costerton, J. W. (2007). In J. W. Costerton (Ed.), *The Biofilm Primer* (pp. 5–13). Springer-Verlage Berlin Heidelberg.
37. Rasmussen, R. M., Epperson, R. T., Taylor, N. B., & Williams, D. L. (2019). Plume height and surface coverage analysis of methicillin-resistant *Staphylococcus aureus* isolates grown in a CDC biofilm reactor. *Biofouling, 35*(4), 463–471.
38. Farha, M. A., & Brown, E. D. (2019). Drug repurposing for antimicrobial discovery. *Nature Microbiology, 4*, 565–577. https://doi.org/10.1038/s41564-019-0357-1.

Biofilm Infections in Orthopedic Surgery and Their Impact on Commercial Product Development

David A. Armbruster

Abstract The biofilm nature of bacterial infections on orthopedic implants imparts antibiotic tolerance, which means that these infections often cannot be successfully treated with systemic antibiotics alone. These infections must instead be treated surgically, which comes with high financial costs and patient morbidity. Because of this, much industrial research and new product development effort has focused on technologies to prevent infection by preventing biofilm formation on implants. Implant surface modification and local antibiotic delivery are primary areas of technology development. Antimicrobial-eluting orthopedic trauma implants have been commercialized and appear to be clinically effective but have not yet enjoyed widespread commercial success. The primary barriers to clinical development of infection-resistant implants in orthopedics are not technical, but commercial and regulatory. The large size and high cost of clinical trials in orthopedics, combined with the fragmented nature of the orthopedic implant market and indication-specific regulatory approvals, make it unlikely that the market for any single implant design can support the cost of clinical data required for approval. The regulatory pathway for any individual product design, typically a function of product risk profile, is a key factor determining clinical data requirements and therefore development costs. By accounting for these non-clinical challenges early in the technology development cycle it may be possible for industry, in partnership with clinicians and regulatory bodies, to bring forward anti-biofilm technologies for orthopedic surgery implants.

Keywords Orthopedic · Infection · Commercial · Implant · Biofilm · Economics · Regulation

D. A. Armbruster (✉)
DePuy Synthes Biomaterials R&D, West Chester, PA, USA
e-mail: darmbrus@its.jnj.com

D. L. Williams (ed.), *Targeting Biofilms in Translational Research, Device Development, and Industrial Sectors*,
https://doi.org/10.1007/978-3-030-30667-0_2

Introduction

The successful clinical development of antibiotics to treat bacterial infections dates to the mid-twentieth century. With the development of Prontosil by Gerhard Domagk in Germany in 1935 and the demonstration of in vivo activity of penicillin by Chain and Florey in 1940, the modern antibiotic era began [1, 2]. Bacterial diseases which had previously claimed the lives of otherwise healthy individuals at a tragic rate were now successfully treated with a 7-day dose of oral or IV antibiotics [3, 4]. Today, uncomplicated staphylococcal or streptococcal skin infections or bloodstream infections are routinely treated with oral antibiotics; however, treatment is significantly complicated in the case of antibiotic-resistant organisms. Resistance to penicillin was documented as early as 1947 [5], and an antibiotic arms race has ensued, with the development of novel antibacterial drugs only barely outpacing the development of resistant organisms.

Much research has led to an understanding of the genetic basis of antibiotic resistance, and the mechanisms of resistance development by single point mutations or plasmid transfer [6]. Bacterial biofilms formed on medical implants are responsible for a completely different mechanism of antibiotic resistance, however, which is now commonly referred to in the context of biofilms as antibiotic tolerance. Rather than a change in the bacterial genotype, antibiotic tolerance in bacterial biofilms is due to a phenotypic shift, where gene expression is altered, extracellular matrix is produced, and cellular metabolism is slowed. Antibiotic tolerance is therefore already built into the genetic code of every biofilm-producing bacterial strain. This biofilm-mediated tolerance creates significant challenges for clinical treatment of implant-related infections. Systemic chemotherapy is often not effective against biofilm-mediated infection of orthopedic implants, and treatment resembles surgical oncology more than pharmacology.

The Clinical Problem

A majority of orthopedic implant-related infections are biofilm-mediated [7, 8]. The presence of a nonviable foreign body in the surgical wound provides a site for bacterial adhesion and implant colonization. Infection rates in clean, elective orthopedic surgery such as knee arthroplasty range from 0.39% to 2.5% [9, 10], while infection rates following internal fixation of high-risk fractures such as open tibial fractures can be 16% or more [11, 12]. Surgical site infections are grouped into the categories of superficial infections, which can be treated by an appropriate course of systemic antibiotics, and deep infections which require surgical intervention for complete eradication of the biofilm. Options for surgical treatment depend on the maturity of the infection, the extent of tissue damage already evident, and antibiotic sensitivity of the cultured pathogen. The least invasive surgical intervention is irrigation and debridement with implant retention; however, this technique is only effective in patients with suspected early infection [13, 14]. For a select subset of relatively healthy patients where the infection is less advanced and the infecting organism is known, a one-stage revision may be appropriate [15]. The infected implant and tissues are removed, and a new implant is put in place at the same time. However in the USA the "gold standard" treatment for infected hip or knee prostheses is a two-stage revision with antibiotic-eluting spacer [16]. Clearly one of the challenges in treating orthopedic implant-related infection is diagnosis, since earlier diagnosis of an infection allows for more conservative treatment.

Implant-related infection can double the overall cost of treatment of orthopedic trauma [17], and treatment cost of an infected knee or hip implant can be five times the cost of a non-infected implant [18]. The more extensive the treatment, the more expensive, so there is pressure to shift toward less extensive treatment options. Certainly there are pathogenic mechanisms other than biofilm formation that allow bacteria to survive surgical debridement and antibiotic treatment, including intracellularization in osteoblasts or fibroblasts [19], and sequestration in the small canaliculi in bone [20]. Together these lead to a high risk of failure of even the most careful and aggressive revision surgery.

The failure of systemic antibiotics to prevent all infections in elective surgery may also be related to the biofilm nature of the bacteria originally contaminating the wound site. The majority of bacteria found in nature, including on human skin, exist as slow-growing mature biofilms. It is likely that even in a "clean" elective surgery, a small number of biofilm-embedded bacteria will contaminate the wound, originating potentially from the wound edge, or as airborne particulates. This biofilm inoculum will already be in an antibiotic-tolerant state due to its low metabolism and extracellular matrix, and even in low numbers of 10^2 or 10^3 colony forming units (CFU) may tolerate systemic antibiotics long enough to form a new nidus of infection [21].

One strategy to kill bacteria in biofilms that have become antibiotic tolerant is to deliver antibiotics directly to the surgical site to create very high local levels. Multiple publications have described the use of local antibiotics placed directly in the surgical wound site to supplement systemic antibiotics. Originally described in spine surgery [22, 23] and more recently for orthopedic trauma, powdered antibiotics such as vancomycin can be sprinkled directly into a wound site [24, 25], mixed with sterile saline into a putty that can be placed onto the orthopedic implant prior to wound closure [26], or injected as a solution into the deep tissue layers following wound closure [27]. Although these studies include small numbers of patients, they have shown trends toward effectiveness in reducing infection rates. To give a more controlled release rate for antibiotic local delivery, antibiotics can be mixed into calcium sulfate bone void fillers. These implant materials have been adapted for use by clinicians in need of more sustained local antibiotic coverage, but the clinical safety and effectiveness is not clear [28].

Impact on Clinical Care and Innovation in Implant Design

General Strategies for Infection Prevention

Given the high cost of treatment for orthopedic implant-related infection, strategies for infection prevention rather than treatment seem to make economic sense. It is hoped that the cost of infection prevention for all patients receiving implants can be more than offset by the savings of treatment costs avoided. The most important tool for infection prevention during any surgery is perioperative systemic antibiotic prophylaxis, and clinical research has demonstrated reduced infection rates [29, 30]. General improvements to the cleanliness of the operating room environment, and "patient optimization" strategies such as chlorhexidine wash for skin decontamination or nasal decontamination, have resulted in lower infection rates for elective surgery [31]. These strategies have brought infection rates down to relatively low levels for elective joint replacement surgeries; however, the high volume of these procedures means that even a low infection rate will impact tens of thousands of joint replacement patients annually and remains a clinical problem [32].

The Importance of Diagnosis

Early diagnosis of orthopedic implant-related infection is important for providing appropriate and timely treatment. Diagnosis is difficult and there is no single test which gives definitive determination of the presence of an infection, due in part to the biofilm nature of implant-related infection [33]. Diagnostic criteria have been developed for prosthetic joint infections which rely on a combination of clinical examination, microbiological cultures and biomarkers of infection [34]. Zimmer Biomet markets the Synovasure® Alpha Defensin Test, originally developed by CD Diagnostics, to aid in the diagnosis of prosthetic joint infection. This is the first diagnostic assay developed primarily for use in diagnosing orthopedic infections due to a broad spectrum of organisms [35, 36]. The original test system was limited in usefulness, since the samples had to be sent off-site for analysis requiring a 24-hour turnaround. A new lateral flow device which provides results within 15 minutes of obtaining a joint aspiration sample has improved time to infection diagnosis, but still does not provide identification of pathogen or antibiotic sensitivity [37].

The use of advanced molecular diagnostics such as polymerase chain reaction electrospray ionization-mass spectrometry (PCR/ESI-MS) or the IBIS 5000 from Abbott Diagnostics have shown promising results in clinical trials for rapid infection diagnosis with pathogen identification [38]. A group at the Mayo Clinic evaluated PCR-ESI/MS for diagnosis of bacterial infection in sonicate fluid from explanted knee or hip prostheses and found that it is more sensitive but less specific than culture for PJI diagnosis [39]. The technique may be useful as an adjunctive method for select cases of arthroplasty failure, but has not proven reliable enough to become a standard method [40]. The clinical need remains for a rapid, low-cost, real-time assay for infection diagnosis and pathogen identification as a tool for guiding treatment of suspected orthopedic implant-related infection.

Infection-Resistant Implants

The first anti-infection technologies brought to the market in orthopedics were drug-eluting bone cement formulations intended for use in revision of infected implants. Surgeons began experimenting with mixing antibiotics into polymethyl-methacrylate (PMMA) bone cement used for arthroplasty procedures in the early 1970s, and cement formulations containing gentamicin were introduced commercially in the USA by 2002 [41]. Molded PMMA spacers are used to maintain the joint space between the two stages of a two-stage hip or knee revision, and spacers containing antibiotics were introduced commercially in the USA by 2003. The designs of these products were based on years of clinical practice of surgeons fashioning their own antibiotic-eluting implants from PMMA intraoperatively and were developed to provide surgeons a more standardized and simplified option. These products helped improve the success rate of two-stage septic revisions of hip and knee prostheses, but PMMA cement used primarily for local drug delivery must eventually be removed in a second surgical procedure. To avoid the need for removal, absorbable calcium-based materials containing antibiotic have been commercialized. Bonesupport, in Lund Sweden, has commercialized antibiotic-containing versions of their Cerament® calcium sulfate/hydroxyapatite synthetic bone substitute. Cerament® V containing 66 mg/mL vancomycin and Cerament® G containing 17.5 mg/mL gentamicin have been used clinically in revision hip arthroplasty procedures [42].

The ultimate solution to the problem of implant-related infection in orthopedic surgery would be to design an implant inherently resistant to bacterial colonization. Preventing bacterial growth and biofilm formation would eliminate this protected niche which provides bacteria with antibiotic tolerance and protection from the immune system. There are two general strategies for making an antibacterial implant: modification of the implant surface to inhibit bacterial growth and local release of an antibacterial agent from the implant surface to kill bacteria before they can attach and colonize the implant. One strategy for implant surface modification relies on chemical interaction of the implant surface to inhibit colonization with bacterial cells, such as in the case of covalently bound antimicrobials. Hickock et al. from Thomas Jefferson University have demonstrated the effectiveness of covalently bound vancomycin to prevent bacterial colonization of metal implants in vivo [43]. Surface modification can also rely on mechanical interaction between bacteria and implant surface features on the micro- or nanoscale. For example, Webster and others have shown an effect of nanoscale surface topography to reduce adhesion and growth of bacteria to TiO_2 surfaces [44, 45].

Antimicrobial-Eluting Implants

Technologies based on local release of antimicrobials from implant surfaces have had some clinical and commercial success. Orthopedic implants designed to elute iodine, silver, or antibiotics such as gentamicin sulfate have demonstrated some clinical efficacy to reduce infection rates [46]. The first antibiotic-eluting orthopedic implants commercialized were poly(methyl methacrylate) (PMMA) bone cements containing antibiotics such as tobramycin or gentamicin. These were developed for use in revision arthroplasty procedures where the risk of infection is significantly higher than primary arthroplasty. PMMA cement with intraoperatively mixed antibiotics has been used for many years for treatment of orthopedic infection, to provide high local antibiotic levels with low risk of systemic toxicity [47].

External fixation pins coated with silver were approved by the FDA in 1996; however, the first commercially successful orthopedic implants with antimicrobial-eluting surface coatings were silver-coated megaprostheses, first sold commercially in Europe in 2002. Megaprostheses (or tumor endoprostheses) are large orthopedic implants used for reconstruction of large bone defects, often following tumor resection or prior implant failure. Due to the size of these implants, the large soft tissue incisions required for their implantation, and the compromised health of the patients in which they are often used, megaprostheses are prone to high rates of infection [48]. Several implant manufacturers have commercialized silver-coated megaprostheses, and multiple clinical case series have been published showing significantly lower rates of implant-related infection as compared to uncoated prostheses [49]. Implantcast in Buxtehude Germany provides an extensive line of their MUTARS® endoprosthesis implants and components with a silver-coated option, and Stanmore Implants, now part of Stryker, has commercialized the Agluna coating on its METS modular tumor system.

The first commercial orthopedic implant coated with an antibiotic drug was the UTN PROtect® tibial nail, first marketed by Synthes in Europe in 2005. This titanium trauma nail used for intramedullary fixation of tibial shaft fractures was coated with a matrix of absorbable poly(D,L-lactide) polymer containing particles of gentamicin sulfate. The gentamicin content of each nail was up to 50 mg, depending on nail diameter and length [50]. This product was followed by the launch of an improved nail design with the same antibiotic coating, the Expert Tibial Nail PROtect®. A clinical case series on the ETN PROtect® nail showed an encouraging safety and efficacy profile in 100 patients [51]. Additional antibiotic-eluting orthopedic implants have been commercialized since 2005 in the EU, such as the CiproScrew™ from Bioretec in Tampere, Finland. This is an absorbable bone screw made from absorbable poly(L-lactide-co-glycolide) containing ciprofloxacin, which is released over approximately 12 weeks as the screw is absorbed.

To date however, antibiotic-coated orthopedic implants have only had limited commercial success, and the technologies have not been widely applied to a broad selection of products. New technologies could be developed to target the biofilm phenotype, and likely improve clinical outcomes even further, but the reason for limited commercial success is not due solely to clinical efficacy. Rather, the complicated and expensive regulatory pathway for bringing these technologies to market is an important limitation.

Regulatory and Commercial Impact of Infection Prevention vs. Treatment Strategies

Years of preclinical research have provided evidence that combining a proven antibiotic drug with an orthopedic implant can be very effective in the majority of cases at reducing implant-related infections. Clinical research has confirmed that local delivery of antibiotics to an orthopedic surgical site can result in a significant reduction in implant-related infection compared to the use of systemic antibiotics alone in high-risk surgeries such as septic revision for total knee implants. However, combining a drug with a medical device into one product for the purpose of preventing implant-related infection comes with many non-technical challenges.

FDA Requirements

In the USA, regulatory approval of new medical devices can be done either via the 510(k) process, based on demonstration of substantial equivalence to a predicate device, or via a Premarket Approval Application (PMA) which requires clinical evidence of device safety and efficacy. The 510(k) process is significantly less costly than the PMA process in terms of both time and money. In 2007 a draft guidance was issued by the FDA outlining its policy on "Premarket Notification (510(k)) Submissions for Medical Devices that Include Antimicrobial Agents." This guidance stated that the 510(k) route is appropriate for a device including an antimicrobial if it incorporates the "same device design and the same antimicrobial agent for the same indication for use" as a predicate device and that "FDA believes the indication for use, 'reduce or prevent device-related infections' should be supported by clinical data." This guidance was subsequently withdrawn by FDA in 2018; however, what this meant was that since 2007, novel antibiotic-coated implants in the USA for which there was no predicate device were required to go through the PMA pathway and human clinical data was required to support any claim of infection reduction. No new antimicrobial-coated products were developed for the orthopedic market during this time, since no predicates exist. Since the 2007 draft guidance was withdrawn, there is no specific guidance from FDA on the regulatory pathway for medical devices modified to prevent biofilm formation.

One additional challenge in commercializing anti-biofilm technologies is that the presence of a biofilm is not a clinical diagnosis. The biofilm theory of implant-related infection assumes that bacterial colonization of an implant surface leads to an implant-related infection. However, direct identification of a bacterial biofilm on an implant in situ is currently not possible in the clinic. Implant-related infection is diagnosed by a combination of clinical signs, such as redness and swelling, serum markers such as CRP, and most importantly positive cultures of wound or synovial fluid [52]. The presence of a bacterial biofilm cannot be confirmed until an implant is removed and examined microbiologically [53]. During diagnosis, the biofilm remains invisible. From a regulatory perspective, this means that a manufacturer has two options with regards to regulatory claims for a new anti-biofilm technology. They may seek to claim a reduction in clinical infection risk, but this must be based on human clinical data. Alternately they may seek to claim prevention of implant colonization or biofilm formation, but this data must necessarily come from a preclinical animal model of infection.

The Economics of Clinical Trials

It is the requirement for human clinical data that creates the most significant hurdle for the infection prevention strategy and specifically for orthopedic drug-device combination products. A clinical trial to show statistically significant infection prevention in orthopedic surgery will require many more patients than a similar trial to show eradication of existing infection. This is due to the relatively low underlying infection rates in elective orthopedic surgery. As an example of a new drug trial for infection treatment, a Phase 3 clinical trial supporting FDA approval of the new antibiotic dalbavancin in 2014 showed noninferiority to vancomycin and linezolid in treatment of skin and skin structure infections. This trial required 568 patients, 284 in each of the treatment and control groups, and was powered at 90% to show a positive response of 85% of patients [54]. In contrast, imagine a novel antibiotic-coated hip prosthesis which is designed to release the drug locally after implantation, reducing infections in primary hip arthroplasty by 50%. A prospective, randomized, controlled clinical trial of this product designed to show a reduction in the baseline 1% infection rate to 0.5% would require over 10,000 patients [55].

The outlook is somewhat better for orthopedic indications with higher underlying infection rates, such as high-energy trauma. Imagine the same antibiotic coating applied to a titanium intramedullary nail used for fixation of tibial fractures and a clinical trial designed to include only patients with severe trauma who are at high risk for infection. In this patient population a prospective, randomized, controlled clinical trial designed to reduce a 12% infection rate to 6% would require close to 1000 patients. The logistics of performing this clinical trial would certainly be more manageable, but the cost may still be prohibitive. A clinical trial of this type can cost in the range of $18,000 per patient, including 1 year of follow-up with all required testing. This proposed trial would take 4–5 years and cost over $18 million, and total product development costs during this time could easily reach $25 million. For a major global medical device company, this cost may not seem out of line; however, an important complicating factor to consider is the indication-specific nature of device approvals in the USA.

The Commercial Challenge

For medical devices in the USA, FDA 510(k) clearance or PMA approval is indication-specific. This means that each product type will require a separate approval and likely separate clinical trials. For example, if a novel anti-infection technology is developed and used to coat an intramedullary nail, a clinical trial would be required to approve this product with a claim of reduced infection rate relative to uncoated nails. If the same coating technology is subsequently applied to a plate and screw system used for fixation of mid-foot fractures, a new clinical trial would be required for approval in this indication.

The fragmentation of the orthopedic market into many different implant types means that the total market for any one type of implant type is limited. As an example, intramedullary tibial nails are one of the highest sales volume implants used in orthopedic trauma surgery, with over 70,000 tibia fractures fixed using tibial nails each year in the USA. Approximately 25% of these are open or high-energy fractures at increased risk for infection, meaning the total US market for an antibiotic-coated tibial nail would be about 17,500 units per year. Since the cost of a clinical trial required for device approval must be spread out over the cost of the product, to make this product worth the investment a manufacturer would need to charge over $10,000 per coated tibial nail. To put this in perspective, Medicare reimbursement for this surgery in the USA is approximately $4500. Clearly, the market forces that drive the business case around an anti-infective implant present a significant barrier to commercialization, even for technologies with strong preclinical evidence for safety and efficacy. The high cost of medical device clinical trials combined with a fragmented market can make commercial development impractical, despite a clear unmet clinical need.

The clinical value of a product based on an anti-biofilm strategy is based on its ability to reduce the overall rate of infection. Only if preclinical data translates into clinical infection reduction is there value in the technology. However, demonstrating this is complicated by the fact that bacteria have various other strategies for survival in bone. For example, bacteria can form small colony variants (SCVs) as a result of genetic mutations in energy metabolic pathways. The slower-growing SCV variants become significantly less susceptible to aminoglycoside antibiotics and can persist in the body for a long period of time until a reverse mutation restores the rapidly growing virulent form [56]. *S. aureus* can sequester themselves into tiny canaliculi in the bone, less than one micron in diameter, where they are out of reach of the immune system [20]. *S. aureus* can even hide inside osteoblasts to evade the immune system and establish a chronic infection [19]. These alternate mechanisms of persistence mean that even though a technology may prevent biofilm formation on an implant, it may not prevent all persistent bone infections at surgical sites. This complicates the task of generating data showing efficacy based on a clinical infection endpoint.

Technical Solutions to a Non-technical Problem

Is it possible to find technical solutions to the commercial and regulatory hurdles for orthopedic implant development with anti-infection properties if these challenges are known from the start? Certainly, the choice of implant and indication will significantly affect clinical trial size and cost. By targeting indications with high infection risk and high procedure volume, the required clinical trial can be smaller in size and can be conducted in a shorter time. Both of these factor into reduced trial costs.

It may also be possible to develop "platform" technologies which can be used with various implants in multiple indications. A single clinical trial may potentially include multiple different implant types and indications, spreading the trial costs over a larger market. The costs of product development would be spread over multiple indications and markets as well. One example of an anti-infection platform technology for orthopedics is the DAC® antibacterial bioabsorbable hydrogel, marketed in Europe by Novagenit. This hyaluronate-polylactide gel is reconstituted intraoperatively from powder using an antibiotic solution of the surgeon's choice and can be used to coat many different types of orthopedic implants. Clinical trials have shown that DAC® gel is effective at reducing infection rates with both trauma implants and joint prostheses [57, 58]. However, a platform technology that functions as a local drug delivery matrix independent of an implantable device would likely be regulated as a pharmaceutical product.

The regulatory pathway for pharmaceuticals is typically more costly in terms of both time and financial investment than that for medical devices, which may negate any advantage due to broader indication for the platform technology. The primary hurdle to drug product development is the cost of two phase 3 clinical trials required for approval in the USA. There may be strategies to lower this barrier, such as negotiating with the FDA for a single clinical trial for products using established generic antibiotics. The commercial market for a local delivery antibiotic formulation intended for use in orthopedics is relatively small, and it may be that no single private company would take on the required risk for this small reward. Successful commercialization of such a product may require public-private partnerships leveraging government funding sources to reduce commercial risk in this space.

Another strategy that may avoid high clinical development costs would be to develop an anti-infection technology based on passive surface modification that does not rely on an antimicrobial agent. The FDA has historically required clinical safety data for implants that include antimicrobial agents, specifically substances that act inside the body to kill or inhibit the growth of microorganisms, and that may result in the emergence of resistance to the antimicrobial agent [59]. As discussed previously, alternate technologies have been explored which rely on surface topography, surface chemistry, or other non-pharmaceutical methods to reduce the tendency for bacteria to attach to an implant surface and form a biofilm. These represent a lower risk to the patient for both systemic toxicity and antibiotic resistance development. As an example, research has shown that introduction of nanoscale roughness to a metal surface by shot peening or by the growth of titanium oxide (TiO_2) nanotubes on anodized titanium surfaces can reduce bacterial adhesion to a surface, as a function of microroughness scale [60, 61]. Nanoscale surface modification can also affect surface energy in a way that alters protein adsorption, allowing further control of the bacteria-biomaterial interaction [62]. Technologies of this type are also attractive commercially because the durable surfaces they produce may be produced with relatively minor modifications of standard manufacturing process. However, technologies based on surface topography have not shown the same degree of effectiveness as antimicrobial-based technologies, and it remains to be shown whether these could be clinically successful.

Risk-Benefit Determination and Regulatory Pathway

It is useful to briefly discuss the importance of risk-benefit determination in the regulatory pathway for anti-infection technologies, especially regarding those that are drug-device combination products. The complexity and cost of regulatory approval for a medical device depends on the regulatory pathway, which is in turn determined primarily by the risk-benefit profile of the device. In the USA, devices are classified by FDA as class I, II, or III, primarily on the basis of their associated risk and the regulatory controls necessary to ensure their safety and effectiveness. Class I devices fall under general controls, and pose the lowest risk, class II require general and specific controls. Class I and II devices are typically approved by the 510(k) process in the USA. Class III devices pose the highest risk and require a PMA for approval in the USA.

Historically, new types of devices which have not been previously classified by FDA have been classified into class III automatically, regardless of the level of risk they pose. This is relevant because many anti-infection technologies for medical devices are novel and have not been previously classified and therefore fall into this category. It is up to the device manufacturer to sufficiently understand all of the probable risks and benefits of the device and how device safety and effectiveness can be assured through the application of general controls or general and special controls [63]. In many cases the PMA path required for approval of a class III product and the attendant requirement for clinical data make the business case impractical for these novel products.

The De Novo 510(k) Process

FDA's de novo 510(k) process provides a pathway to Class I or Class II classification for moderate-risk medical devices for which general controls or general and special controls would provide a reasonable assurance of safety and effectiveness, but for which there is no legally marketed predicate device. A de novo submission can be made if a 510(k) submission receives a determination of "not substantially equivalent," or as a "direct de novo" submission without a prior 510(k) submission if the manufacturer believes the device is appropriate for classification into Class I or Class II and there is no legally marketed predicate device. For manufacturers seeking FDA approval for drug-device combination products, the choice of active agent may play a significant role in allowing a de novo submission. Many active agents used in drug-device combination products are not novel drug products, but well-characterized generic drugs or antimicrobials. If the sponsor believes that the combination of a well-established Class II medical device with an active agent that has a proven safety profile is low risk and can be regulated by general and special controls, a de novo 510(k) application may be valid.

In its guidance document on making risk-benefit determinations for medical device submissions, FDA outlines multiple factors that should be addressed. A number of those are especially relevant to anti-infection technologies. The type, magnitude, and duration of the benefit and the patient's perception of the severity of the disease must be taken into account. On the risk side, the severity, types, number, and rates of harmful events as well as their probability and duration are factors. Importantly, the availability of alternate treatments and the novelty of the technology for addressing an unmet medical need are important factors to understand. In assessing benefit and risk, FDA considers whether a device is a breakthrough technology that addresses an unmet medical need, such as providing a treatment where no alternative is available [63]. This may be relevant for some anti-infection technologies, especially those intended for high-risk indications where infection is a significant threat.

Regulatory Engagement: FDA

The US Food and Drug Administration recognizes that the lack of regulatory guidance and lack of standardized testing for anti-biofilm technologies make regulatory review more complicated and are open to engagement with industry to help create more consistency. For medical devices containing antimicrobial drugs, the recommended first point of contact is FDA's Office of Combination Products (OCP). A Request for Designation (RFD) can be submitted to OCP to determine the primary mode of action for a drug-device combination product, and OCP coordinates reviews with multiple agencies to ensure timeliness and consistency.

FDA has also taken steps to encourage biofilm research and industry engagement. FDA scientists and regulators present regularly at biomaterials and biofilm conferences, such as those sponsored by the Center for Biofilm Engineering at Montana State University [64]. FDA's Center for Devices and Radiological Health (CDRH) conducts its own research to advance regulatory science via the Office of Science and Engineering Laboratories (OSEL). Within OSEL, The Division of Biology, Chemistry, and Materials Science (DBCMS) has a group focused on Microbiology and Infection Control, which studies biofilm diagnostics and detection as well as bacterial interaction with medical device materials [65]. FDA's National

Center for Toxicological Research also has a Division of Microbiology, which studies *S. aureus* biofilms grown in bioreactors on medical device materials [66]. With continued industry-FDA engagement, hopefully more expedited pathways will emerge for the commercialization of anti-biofilm technologies.

Summary

Bacterial biofilm formation on orthopedic implants and the resulting antibiotic tolerance cause significant problems for the treatment of orthopedic implant-related infections. Instead of a short course of oral or intravenous antibiotics, implant-related infections often must be treated by surgical removal, debridement of infected tissue, and extended local and systemic antibiotic treatment, followed by additional surgery for replacement of the implant. The cost and morbidity associated with surgical treatment of infections have therefore caused researchers and implant manufacturers to focus on technologies for infection prevention.

Many viable anti-infection technologies have been demonstrated preclinically, and several of these have obtained regulatory approval; however, despite some clinical success, these products have not seen widespread commercial success. Significant non-technical barriers exist to commercial development of infection prevention strategies in orthopedics. The requirement for clinical data to demonstrate safety and efficacy for a novel technology in the USA adds significant cost and time to product development. The fragmentation of the orthopedic device market into many specialized implants, combined with indication-specific regulatory approvals, means that the market for any one product may not be large enough to bear the high development costs required.

By factoring the non-technical challenges into the development of anti-infection technologies from the beginning, it may be possible to identify technical solutions to these non-technical problems. Regulatory bodies should also consider requirements and guidelines being put in place, and whether they will allow technology innovation to match healthcare needs. There is additional room for industry and regulatory bodies to meet in the middle and to provide clearer pathways and even incentives for development of novel anti-infection products. A thorough understanding of regulatory requirements and commercial considerations is essential, and close collaboration with regulatory bodies from the start of a development program can help navigate a viable route.

References

1. Hörlein, H. (1935). The chemotherapy of infectious diseases caused by Protozoa and Bacteria. *Proceedings of the Royal Society of Medicine, XXIX*(313).
2. Chain, E., Florey, H. W., Gardner, A. D., et al. (1940). Penicillin as a chemotherapeutic agent. *The Lancet, 1940*, 226–228.

3. Foulis, M. A., & Barr, J. B. (1937). Prontosil album in puerperal sepsis. *British Medical Journal, 27*(1937), 445–446.
4. Abraham, E. P., Chain, E., Fletcher, C. M., et al. (1941). Further observations on penicillin. *The Lancet, 1941*, 177–188.
5. Barber, M. (1947). Staphylococcal infection due to penicillin-resistant strains. *British Medical Journal, 2*(4534), 863–865.
6. Alekshun, M. N., & Levy, S. B. (2007). Molecular mechanisms of antibacterial multidrug resistance. *Cell, 128*(6), 1037–1050.
7. Stoodley, P., Ehrlich, G. D., Sedghizadeh, P. P., et al. (2011). Orthopaedic biofilm infections. *Current Orthopaedic Practice, 22*(6), 558–563.
8. Nana, A., Nelson, S. B., & McLaren, A. (2016). What's new in musculoskeletal infection: Update on biofilms. *The Journal of Bone and Joint Surgery. American Volume, 98*(14), 1226–1234.
9. Voigt, J., Mosier, M., & Darouiche, R. (2015). Systematic review and meta-analysis of randomized controlled trials of antibiotics and antiseptics for preventing infection in people receiving primary total hip and knee prostheses. *Antimicrobial Agents and Chemotherapy, 59*(11), 6696–6707.
10. Namba, R. S., Inacio, M. C., & Paxton, E. W. (2013). Risk factors associated with deep surgical site infections after primary total knee arthroplasty: An analysis of 56,216 knees. *The Journal of Bone and Joint Surgery. American Volume, 95*(9), 775–782.
11. Stall, A., Paryavi, E., Gupta, R., et al. (2013). Perioperative supplemental oxygen to reduce surgical site infection after open fixation of high-risk fractures: A randomized controlled pilot trial. *Journal of Trauma and Acute Care Surgery, 75*(4), 657–663.
12. Bosse, M. J., Murray, C. K., Carlini, A. R., et al. (2017). Assessment of severe extremity wound bioburden at the time of definitive wound closure or coverage: Correlation with subsequent postclosure deep wound infection (bioburden study). *Journal of Orthopaedic Trauma, 31*(Suppl 1), S3–S9.
13. Urish, K. L., Bullock, A. G., Kreger, A. M., et al. (2018). A multicenter study of irrigation and debridement in Total knee arthroplasty Periprosthetic joint infection: Treatment failure is high. *The Journal of Arthroplasty, 33*(4), 1154–1159.
14. Grammatopoulos, G., Bolduc, M. E., Atkins, B. L., et al. (2017). Functional outcome of debridement, antibiotics and implant retention in periprosthetic joint infection involving the hip: A case-control study. *The Bone & Joint Journal, 99-B*(5), 614–622.
15. Haddad, F. S., Sukeik, M., & Alazzawi, S. (2015). Is single-stage revision according to a strict protocol effective in treatment of chronic knee arthroplasty infections? *Clinical Orthopaedics and Related Research, 473*(1), 8–14.
16. Cooper, H. J., & Della Valle, C. J. (2013). The two-stage standard in revision total hip replacement. *The Bone & Joint Journal, 95-B*(11 Suppl A), 84–87.
17. Thakore, R. V., Greenberg, S. E., Shi, H., et al. (2015). Surgical site infection in orthopedic trauma: A case-control study evaluating risk factors and cost. *Journal of Clinical Orthopaedics & Trauma, 6*(4), 220–226.
18. Stambough, J. B., Nam, D., Warren, D. K., et al. (2017). Decreased hospital costs and surgical site infection incidence with a universal decolonization protocol in primary total joint arthroplasty. *The Journal of Arthroplasty, 32*(3), 728–734.e1.
19. Horn, J., Stelzner, K., Rudel, T., et al. (2017). Inside job: Staphylococcus aureus host-pathogen interactions. *International Journal of Medical Microbiology, S1438-4221*(17), 30316–30318.
20. de Mesy Bentley, K. L., Trombetta, R., Nishitani, K., et al. (2017). Evidence of Staphylococcus Aureus deformation, proliferation, and migration in canaliculi of live cortical bone in murine models of osteomyelitis. *Journal of Bone and Mineral Research, 32*(5), 985–990.
21. Williams, D. L., & Costerton, J. W. (2012). Using biofilms as initial inocula in animal models of biofilm-related infections. *Journal of Biomedical Materials Research. Part B, Applied Biomaterials, 100*(4), 1163–1169.

22. Chiang, H. Y., Herwaldt, L. A., Blevins, A. E., et al. (2014). Effectiveness of local vancomycin powder to decrease surgical site infections: A meta-analysis. *The Spine Journal, 14*(3), 397–407.
23. Evaniew, N., Khan, M., Drew, B., et al. (2015). Intrawound vancomycin to prevent infections after spine surgery: A systematic review and meta-analysis. *European Spine Journal, 24*(3), 533–542.
24. Owen, M. T., Keener, E. M., Hyde, Z. B., et al. (2017). Intraoperative topical antibiotics for infection prophylaxis in pelvic and acetabular surgery. *Journal of Orthopaedic Trauma, 31*(11), 589–594.
25. Singh, K., Bauer, J. M., LaChaud, G. Y., et al. (2015). Surgical site infection in high-energy peri-articular tibia fractures with intra-wound vancomycin powder: A retrospective pilot study. *Journal of Orthopaedics and Traumatology, 16*(4), 287–291.
26. Firoozabadi, R., Miranda, S., & Tornetta, P., 3rd. (2017). Technique for placement of peri-implant antibiotics using antibiotic putty. *Journal of Orthopaedic Trauma, 31*(12), e442–e445.
27. Lawing, C. R., Lin, F. C., & Dahners, L. E. (2015). Local injection of aminoglycosides for prophylaxis against infection in open fractures. *The Journal of Bone and Joint Surgery. American Volume, 97*(22), 1844–1851.
28. Flierl, M. A., Culp, B. M., Okroj, K. T., et al. (2017). Poor outcomes of irrigation and debridement in acute periprosthetic joint infection with antibiotic-impregnated calcium sulfate beads. *The Journal of Arthroplasty, 32*(8), 2505–2507.
29. Boxma, H., Broekhuizen, T., Patka, P., et al. (1996). Randomised controlled trial of single-dose antibiotic prophylaxis in surgical treatment of closed fractures: The Dutch Trauma Trial. *Lancet, 347*(9009), 1133–1137.
30. Bratzler, D. W., Dellinger, E. P., Olsen, K. M., et al. (2013). Clinical practice guidelines for antimicrobial prophylaxis in surgery. *American Journal of Health-System Pharmacy, 70*(3), 195–283.
31. Schweizer, M., Perencevich, E., McDanel, J., et al. (2013). Effectiveness of a bundled intervention of decolonization and prophylaxis to decrease Gram positive surgical site infections after cardiac or orthopedic surgery: Systematic review and meta-analysis. *BMJ, 346*, f2743.
32. Rasouli, M. R., Restrepo, C., Maltenfort, M. G., et al. (2014). Risk factors for surgical site infection following total joint arthroplasty. *The Journal of Bone and Joint Surgery. American Volume, 96*(18), e158.
33. Nelson, C. L., McLaren, A. C., McLaren, S. G., et al. (2005). Is aseptic loosening truly aseptic? *Clinical Orthopaedics and Related Research*, (437), 25–30.
34. Parvizi, J., Zmistowski, B., Berbari, E. F., et al. (2011). New definition for periprosthetic joint infection: From the Workgroup of the Musculoskeletal Infection Society. *Clinical Orthopaedics and Related Research, 469*(11), 2992–2994.
35. Deirmengian, C., Kardos, K., Kilmartin, P., et al. (2015). The alpha-defensin test for periprosthetic joint infection responds to a wide spectrum of organisms. *Clinical Orthopaedics and Related Research, 473*(7), 2229–2235.
36. Bonanzinga, T., Zahar, A., Dütsch, M., et al. (2017). How reliable is the alpha-defensin immunoassay test for diagnosing periprosthetic joint infection? A prospective study. *Clinical Orthopaedics and Related Research, 475*(2), 408–415.
37. Gehrke, T., Lausmann, C., Citak, M., et al. (2018). The accuracy of the alpha defensin lateral flow device for diagnosis of periprosthetic joint infection: Comparison with a gold standard. *The Journal of Bone and Joint Surgery. American Volume, 100*(1), 42–48.
38. Vincent, J. L., Brealey, D., Libert, N., et al. (2015). Rapid diagnosis of infection in the critically ill, a multicenter study of molecular detection in bloodstream infections, pneumonia, and sterile site infections. *Critical Care Medicine, 43*(11), 2283–2291.
39. Greenwood-Quaintance, K. E., Uhl, J. R., Hanssen, A. D., et al. (2014). Diagnosis of prosthetic joint infection by use of PCR-electrospray ionization mass spectrometry. *Journal of Clinical Microbiology, 52*(2), 642–649.

40. Rasouli, M. R., Harandi, A. A., Adeli, B., et al. (2012). Revision total knee arthroplasty: Infection should be ruled out in all cases. *The Journal of Arthroplasty, 27*(6), 1239–43.e1–2.
41. Levin, P. D. (1975). The effectiveness of various antibiotics in methyl methacrylate. *Journal of Bone and Joint Surgery. British Volume (London), 57*(2), 234–237.
42. Logoluso, N., Drago, L., Gallazzi, E., et al. (2016). Calcium-based, antibiotic-loaded bone substitute as an implant coating: A pilot clinical study. *Journal of Bone and Joint Infection, 1*, 59–64.
43. Antoci, V., Jr., Adams, C. S., Hickok, N. J., et al. (2007). Vancomycin bound to Ti rods reduces periprosthetic infection: Preliminary study. *Clinical Orthopaedics and Related Research, 461*, 88–95.
44. Liu, L., Bhatia, R., & Webster, T. J. (2017). Atomic layer deposition of nano-TiO_2 thin films with enhanced biocompatibility and antimicrobial activity for orthopedic implants. *International Journal of Nanomedicine, 12*, 8711–8723.
45. Jaggessar, A., Shahali, H., Mathew, A., et al. (2017). Bio-mimicking nano and micro-structured surface fabrication for antibacterial properties in medical implants. *Journal of Nanobiotechnology, 15*(1), 64.
46. Alt, V. (2017). Antimicrobial coated implants in trauma and orthopaedics-A clinical review and risk-benefit analysis. *Injury, 48*(3), 599–607.
47. Wahlig, H., Dingeldein, E., Bergmann, R., et al. (1978). The release of gentamicin from polymethylmethacrylate beads. An experimental and pharmacokinetic study. *Journal of Bone and Joint Surgery. British Volume (London), 60-B*(2), 270–275.
48. Hardes, J., Gebert, C., Schwappach, A., et al. (2006). Characteristics and outcome of infections associated with tumor endoprostheses. *Archives of Orthopaedic and Trauma Surgery, 126*(5), 289–296.
49. Hardes, J., Henrichs, M. P., Hauschild, G., et al. (2017). Silver-coated megaprosthesis of the proximal tibia in patients with sarcoma. *The Journal of Arthroplasty, 32*(7), 2208–2213.
50. Fuchs, T., Stange, R., Schmidmaier, G., et al. (2011). The use of gentamicin-coated nails in the tibia: Preliminary results of a prospective study. *Archives of Orthopaedic and Trauma Surgery, 131*(10), 1419–1425.
51. Schmidmaier, G., Kerstan, M., Schwabe, P., et al. (2017). Clinical experiences in the use of a gentamicin-coated titanium nail in tibia fractures. *Injury, 48*(10), 2235–2241.
52. Parvizi, J., Tan, T. L., Goswami, K., et al. (2018). The 2018 definition of periprosthetic hip and knee infection: An evidence-based and validated criteria. *The Journal of Arthroplasty, 33*(5), 1309–1314.e2.
53. Stoodley, P., Nistico, L., Johnson, S., et al. (2008). (2008) Direct demonstration of viable Staphylococcus aureus biofilms in an infected total joint arthroplasty. A case report. *The Journal of Bone and Joint Surgery. American Volume, 90*(8), 1751–1758.
54. Boucher, H. W., Wilcox, M., Talbot, G. H., et al. (2017). Once-weekly dalbavancin versus daily conventional therapy for skin infection. *The New England Journal of Medicine, 370*(23), 2169–2179.
55. Sealed Envelope Ltd. (2012). Power calculator for binary outcome superiority trial. https://www.sealedenvelope.com/power/binary-superiority/
56. Sendi, P., Rohrbach, M., Graber, P., et al. (2006). Staphylococcus aureus small colony variants in prosthetic joint infection. *Clinical Infectious Diseases, 43*(8), 961–967.
57. Romanò, C. L., Malizos, K., Capuano, N., et al. (2016). Does an antibiotic-loaded hydrogel coating reduce early post-surgical infection after joint arthroplasty? *Journal of Bone and Joint Infection, 1*, 34–41.
58. Malizos, K., Blauth, M., Danita, A., et al. (2017). Fast-resorbable antibiotic-loaded hydrogel coating to reduce post-surgical infection after internal osteosynthesis: A multicenter randomized controlled trial. *Journal of Orthopaedics and Traumatology, 18*(2), 159–169.
59. Food and Drug Administration Center for Devices and Radiological Health. (2007). Draft guidance for industry and FDA staff – premarket notification [510(k)] submissions for medical devices that include antimicrobial agents. Retrieved from https://wayback.archive-it.org/7993/20180423155127/https://www.fda.gov/MedicalDevices/DeviceRegulationandGuidance/GuidanceDocuments/ucm071380.htm

60. Kummer, K. M., Taylor, E. N., Durmas, N. G., et al. (2013). Effects of different sterilization techniques and varying anodized TiO_2 nanotube dimensions on bacteria growth. *Journal of Biomedical Materials Research. Part B, Applied Biomaterials, 101*(5), 677–688.
61. Bagherifard, S., Hickey, D. J., de Luca, A. C., et al. (2015). The influence of nanostructured features on bacterial adhesion and bone cell functions on severely shot peened 316L stainless steel. *Biomaterials, 73*, 185–197.
62. Bhardwaj, G., & Webster, T. J. (2017). Reduced bacterial growth and increased osteoblast proliferation on titanium with a nanophase TiO2 surface treatment. *International Journal of Nanomedicine, 12*, 363–369.
63. Food and Drug Administration Center for Devices and Radiological Health. (2016). Guidance for industry and food and drug administration staff - Factors to consider when making benefit-risk determinations in medical device premarket approval and de novo classifications. Retrieved from https://www.fda.gov/downloads/medicaldevices/deviceregulationandguidance/guidancedocuments/ucm517504.pdf
64. Phillips, K. S. (2018). What can we learn about medical device associated infection pathogenesis from skin explant models? Presentation at CBE Conference on Anti-Biofilm Technologies: Pathways to Product Development, Arlington VA, 7 Feb 2018.
65. From FDA web site: https://www.fda.gov/MedicalDevices/ScienceandResearch/ResearchPrograms/ucm477398.htm
66. From FDA web site: https://www.fda.gov/AboutFDA/CentersOffices/OC/OfficeofScientificandMedicalPrograms/NCTR/WhatWeDo/ResearchDivisions/ucm079048.htm

Catheter-Associated Urinary Tract Infections: Development of a Test Method for Assessing the Efficacy of Antimicrobial Technologies/Products

Jennifer Summers and Darla M. Goeres

Abstract Urinary catheters are one of the most commonly utilized medical devices worldwide. They are used in virtually every healthcare setting and contribute to improvements in patient care. While urinary catheters provide invaluable aid to patients, they are not without complications, the most notable being catheter-associated urinary tract infections (CAUTI). Once a urinary catheter is in place, pathogens may migrate to the bladder one of two ways, through the catheter lumen or extraluminally in the periurethral space. In vitro methods are useful tools for predicting clinical efficacy only if they accurately model the most important factors contributing to a clinical infection. Although in vitro methods can't replace in vivo scenarios, outcomes may be improved as experiments approximate relevant criteria including the biofilm phenotype, time, media type, materials similarities and environment.

Keywords Urinary catheter · Biofilm · Infection · In vitro · Model · Relevant · Repeatable

Introduction

Urinary catheters are one of the most commonly utilized medical devices worldwide. They are used in virtually every healthcare setting and contribute to improvements in patient care by relieving urinary retention, reducing risk for injury following traumatic surgery, and allowing for accurate urine output readings (e.g.,

J. Summers
Center for Biofilm Engineering, Montana State University, Bozeman, MT, USA

D. M. Goeres (✉)
Center for Biofilm Engineering, Department of Chemical & Biological Engineering, Montana State University, Bozeman, MT, USA
e-mail: darla_g@montana.edu

D. L. Williams (ed.), *Targeting Biofilms in Translational Research, Device Development, and Industrial Sectors*,
https://doi.org/10.1007/978-3-030-30667-0_3

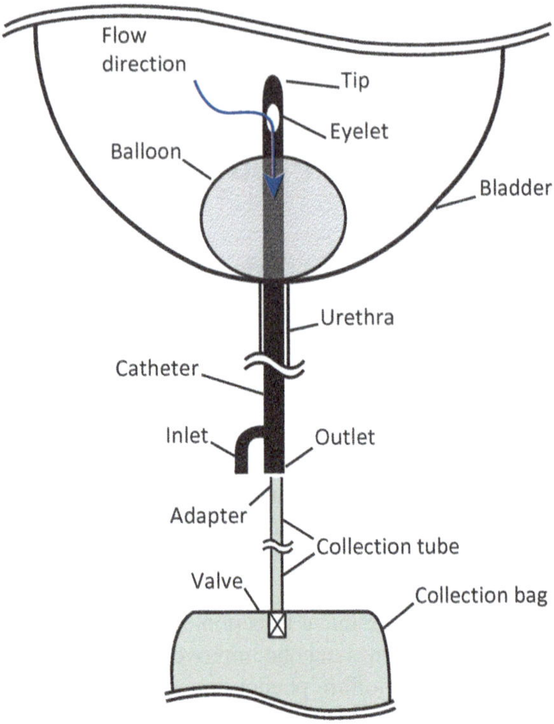

Fig. 1 Diagram of two-way Foley catheter inserted through urethra into the bladder. The blue arrow represents the flow of urine through the catheter into the external collection bag. (Reprinted from reference [4] with permission from publisher)

hemodynamics, electrolyte balance). The global market for urinary catheters is expected to grow to $5.51 billion by 2024,[1] largely due to the increasing elderly, obese, and diabetic populations. The most common urinary catheter is the Foley catheter, which consists of a tube that is inserted through the urethra and is held in the bladder by an inflatable balloon (Fig. 1). While urinary catheters provide invaluable aid to patients, they are not without complications, the most notable being catheter-associated urinary tract infections (CAUTI). The CDC defines CAUTI as a positive urine culture ($\geq 10^5$ colony forming unit (CFU) per milliliter (mL)) in concordance with at least one common symptom (e.g., fever, urinary urgency, and frequency) in a patient who has had a catheter in place for longer than 2 days [1]. Catheter-associated infections account for 37% of all hospital-acquired infections (HAI) and 70% of all nosocomial urinary tract infections (UTI) in the USA [2, 3].

The last major advancement in reducing CAUTI rates was with the introduction of the closed drainage system in the late 1950s (Fig. 2). This improvement resulted in a 50% reduction of incidence of infection [8, 9]. Continued advancements were made in aseptic catheterization techniques in the decades following, but infection rates didn't fall below 30% [10]. Garibaldi et al. hypothesized that with the use of

[1] https://www.grandviewresearch.com/industry-analysis/urinary-catheters-market. Accessed June 19, 2018.

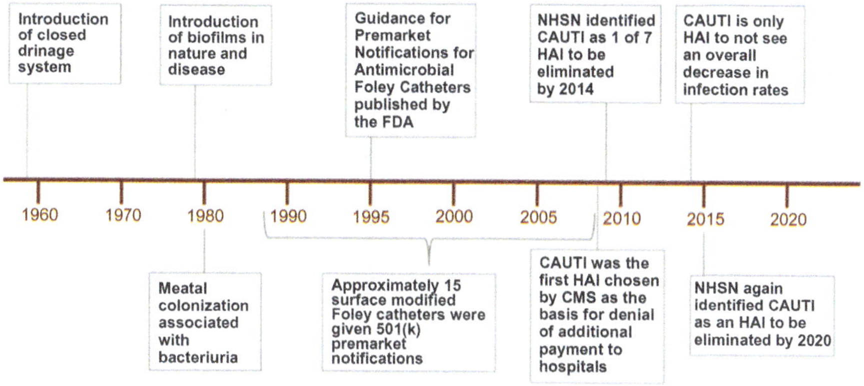

Fig. 2 Timeline showing the evolution of urinary catheters. The modern Foley catheter was invented by Dr. Frederic Foley in 1937 [5]. However, the use of tubes, reeds, straws, etc. to relieve urinary retention has been recorded since the time of the early Egyptians, approximately 1500 BC [6, 7]

aseptic closed drainage systems, the point of entry for bacteria into the bladder became the periurethral mucous sheath outside the catheter [11]. While this hypothesis was eventually proved correct, researchers were still missing a piece of the puzzle. In the late 1970s Dr. William (Bill) Costerton introduced the missing piece with his hypothesis of how bacteria stick to surfaces. He suggested bacteria colonize a surface by producing a glycocalyx of fibers which adhere to a surface and aggregates of cells form complex colonies, or biofilm [12]. This discovery led to many investigative studies throughout the 1980s all inquiring what role biofilm plays in medical infections [13, 14]. By the end of the decade, researchers had found that bacteria form biofilms on almost any surface, from rocks in streams to intrauterine devices (IUDs) [15]. Since, it has been shown that virtually all implantable medical devices are susceptible to biofilm colonization and infection [16–20].

Following this advancement, the FDA published Premarket Notifications for Antimicrobial Foley catheters in 1995 to help industry develop new technologies [21]. 510(k) approvals were given to 15 surface-modified catheters between 1987 and 2008.[2] However, even with the introduction of these new technologies, CAUTI still accounted for approximately 30% of all nosocomial infections [22]. Due to the significant economic and labor-intensive burden of these infections, the Center for Medicaid and Medicare (CMM) identified CAUTI as the first hospital-associated infection that was the basis for denial of government aid to hospitals in 2008 [23, 24]. In response, the CDC published the "National Action Plan to Prevent Health Care-Associated Infections: Road Map to Elimination (HAI Action Plan)" the following year [25]. The plan identified the top seven HAIs reported in the USA, including CAUTI, and set incidence rate reduction goals for each. Despite

[2] https://www.accessdata.fda.gov/scripts/cdrh/cfdocs/cfpmn/pmn.cfm. Accessed June 19, 2018.

advancements in understanding the infection pathogenesis and surface modification technology, CAUTI was the only HAI to lack a significant decrease in incidence rate between 2009 and 2014 [22].

This chapter will explore why CAUTI is challenging to resolve through an examination of the current understanding of the pathogenesis and the necessary attributes and limitations of in vitro models used to evaluate the surface-modified catheters developed to prevent it.

Mechanisms of Infection

To properly assess whether a surface-modified catheter will perform as designed a robust understanding of the pathogenesis of CAUTI is necessary. Clinically, UTI are classified as either uncomplicated or complicated. Uncomplicated UTIs occur in otherwise healthy individuals with no urinary tract abnormalities [26]. Typically, the body's innate defense mechanisms, such as micturition, mucosal secretions, and urine itself, make it more challenging for an infection to start [27–30]. However, there are several risk factors associated with uncomplicated UTIs, including sexual activity, the female gender, and diabetes [31–33]. Complicated UTIs are a result of factors which compromise the body's natural defenses, such as urinary retention from neurological diseases or the presence of a foreign object [34, 35]. Upon insertion of a foreign object, such as a urinary catheter, most of the host defenses are disrupted, and a patient is significantly more susceptible to infection [36, 37]. The CDC advises that each day an indwelling urinary catheter remains in situ, a patient has a 3–7% increased risk for infection [1, 38]. Many hospital patients are inherently immunodeficient. Further compromising a patient's natural defenses with a catheter allows them to be easily colonized by cross-transmission through the hands of healthcare professionals or even by their own perineal flora [13, 39, 40]. Once a urinary catheter is in place, pathogens may migrate to the bladder one of two ways, through the catheter lumen or extraluminally in the periurethral space [41, 42]. Intraluminal contamination accounts for approximately 33% of CAUTIs and is most commonly associated with a break in the closed sterile system or contamination of the collection bag urine [42, 43]. In this case, bacteriuria has been shown to occur within 48 hours [44]. If a strict sterile collection system is kept, the extraluminal route becomes more important, accounting for 66% of CAUTIs [42, 45]. Extraluminal contamination may occur early if the tip of the catheter is contaminated upon insertion, which then smears the bacteria evenly up the urethra into the bladder, or later by bacteria ascending the periurethral space outside the catheter [39, 46]. If late contamination occurs, bacteriuria may take as long as 168 hours to become apparent [44]. CAUTIs may be caused by a broad spectrum of bacteria, both gram negative and gram positive, and certain fungi. The most common causative agents include uropathogenic *Escherichia coli* (UPEC), *Candida albicans*, *Enterococcus* sp., and *Pseudomonas aeruginosa* (Table 1) [3].

Table 1 Distribution and rank order of pathogens frequently reported with CAUTIs across the USA for two reporting periods

	January 2006–October 2007		January 2011–December 2014	
Pathogen	Percent of pathogenic isolates	Rank[a]	Percent of pathogenic isolates	Rank[a]
Coagulase-negative staphylococci	2.5	7	2.4	13
Staphylococcus aureus	2.2	8	1.6	14
Enterococcus species		3		
E. faecalis	3.6		7.0	5
E. faecium	6.0		2.7	11
Other *Enterococcus* or NOS	5.3		4.1	7
Candida species		2		
C. albicans	14.5		11.7	2
Candida glabrata			2.7	12
Other *Candida* spp. or NOS	6.5		3.4	10
Escherichia coli	21.4	1	23.9	1
Pseudomonas aeruginosa	10.0	4	10.3	3
Klebsiella spp.			10.1	4
K. pneumoniae	7.7	5		
K. oxytoca	0.9	10		
Enterobacter spp.	4.1	6	3.7	9
Acinetobacter baumannii	1.2	9		
Yeast NOS			6.1	6
Proteus spp.			4.0	8
Other	14.1		6.4	
Total	100		100	

Reporting criteria changed between the two periods
[a]The 10 (2006–2007) and 14 (2011–2014) most common pathogens are listed and ranked according to how frequently they were reported to the CDC's National Health Safety Network (NHSN) [3, 47]. The rankings were established based on all pathogens reported

For a uropathogen to initiate infection, it must first attach to either the uroepithelial tissue or the catheter surface [48, 49]. There is limited knowledge on the specific adhesions pathogens use to adhere to catheter surfaces; however, researchers have made inferences based on the knowledge of pathogenesis during uncomplicated UTIs. For example, it is hypothesized that UPEC employs type 1 fimbriae to adhere to catheter surfaces as it has been observed using fimbriae to attach to and colonize host tissue in uncompromised urinary tracts [37, 40]. In addition to initiating infection, preliminary in vitro studies have shown the expression of type 1 and type 3 fimbriae by *Klebsiella pneumoniae* promotes biofilm formation on the surface of urinary catheters [50]. The formation of biofilm is a common survival tactic employed by uropathogens, allowing them to persist and cause recurrent infections [51, 52]. The biofilm structure shields organisms from the stresses of the harsh environment such as antimicrobial treatment, host immune responses, and urine itself, described in detail elsewhere [52–57]. A classic example of these protective communities formed during CAUTIs is the crystalline biofilms produced by *Proteus* species. *Proteus mirabilis*, along with other uropathogens, produces urease, which hydrolyses urea to produce carbon dioxide and ammonia [52, 58]. This increases the pH of the local environment, subsequently generating calcium crystals and magnesium ammonium phosphate (struvite) precipitates [40, 51]. These stones become entangled with the bacteria (Fig. 3a) and continue to build on one another ultimately resulting in complete blockage of the catheter lumen (Fig. 3b) [59].

The longer a catheter remains in place, the greater the possibility these organisms will form biofilms and the greater the risk for adverse events (i.e., catheter obstruction, kidney infection, etc.). The possibility of CAUTI is so likely that prevention guidelines go as far as to advise healthcare professions to avoid the use of indwelling urinary catheter when possible [24], although this is not a feasible option for many patients. Clinicians are also limited to which antibiotics they may prescribe due to the increasing frequency of antibiotic resistance among pathogens [3, 60], leaving the healthcare industry desperate for realistic treatment options. It would be a considerable benefit for patients to have a technology available which significantly delayed or completely inhibited the bacterial colonization of urinary catheters.

A literature review found there is a critical lack of data that correlates infection rates to the number of biofilm bacteria on a catheter surface. This information is necessary to translate in vitro data, often reported as a log reduction, to the ability of a surface-modified catheter to prevent infection. This chapter will focus on in vitro methods that may be used to assess surface-modified (including antimicrobial) urinary catheters to prevent or reduce biofilm accumulation. A robust understanding of the pathogenesis of CAUTI enables for a better test that accurately depicts the conditions under which the device will be employed.

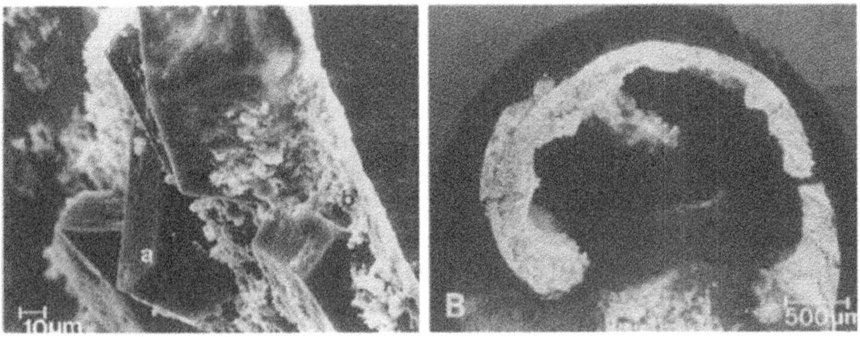

Fig. 3 Freeze-dried preparations of ex vivo catheters colonized by pure cultures of *Proteus mirabilis* showing large crystals (**a**) and a cross section of the blockage in lumen (**b**). (Reprinted from reference [59] with permission from publisher)

Determining How Best to Assess the Safety, Efficacy, Quality, and Performance of Antimicrobial Products with In Vitro Testing

Over the last three decades, researchers have modified the surface of urinary catheters to reduce bacterial attachment, trying everything from impregnating the catheters with antibiotics to printing micropatterns on the surface [61–63]. One of the few modified catheters to make it to market and remain there was a hydrogel-silver-coated catheter. Initial in vitro testing demonstrated hydrogel-silver-coated latex catheters significantly reduced the adhesion of *E.coli* and *P. aeruginosa* when compared to uncoated catheters [64]. This study ultimately led to the product's approval for clinical use in the late 1990s. Since the approval, there have been multiple clinical trials involving hydrogel-silver-coated catheters which have contradicting conclusions. A few of the studies reported favorable results for the antimicrobial catheters, others reported no difference at all, and some found a difference but concluded the difference was not enough for there to be an economic benefit in using the coated catheters [65–68]. So, what happened, why don't the clinical results correlate with in vitro results? A more critical review of in vitro methods used by Gabriel et al. demonstrates the methods do not reflect the conditions existing during an infection, and the data was over extrapolated, resulting in an inaccurate prediction of clinical performance.

The goal is for an in vitro test to predict a device's performance in a clinical setting. To achieve this goal, the in vitro method should model the most important factors contributing to a clinical infection as closely as possible, keeping in mind it is not possible to model host interaction in vitro (i.e., immune response). To further demonstrate the importance of model design, the efficacy of a chlorhexidine-coated urinary catheter against a UPEC was evaluated using three different in vitro methods: zone of inhibition (ZOI), liquid broth test, and a flow-through method. The ZOI and liquid broth test, both classic microbiological assays, were run as described in Table 2. The flow-through method, shown in Fig. 4, incorporates three critical environmental factors seen in vivo: the flow of nutrients through the lumen, use of a relevant growth media (artificial urine media (AUM)), and a time frame relevant to the device.

Results from the ZOI and liquid broth test are shown in Fig. 5. The ZOI for the coated catheters was 11.4 mm ± 2.89. This ZOI is comparable to an antimicrobial catheter currently on the market [69]. The liquid broth test resulted in a 99.9% reduction in bacterial adherence. These results are equivalent to those seen in the C.R. Bard, Inc. sponsored test discussed above, which suggested hydrogel-silver-coated catheters decrease the bacterial adherence of *E.coli* by 99% [64]. However, when the chlorhexidine-coated catheter was evaluated in the flow-through model (Fig. 6) by 24 hours, the surface of the chlorhexidine-coated catheters already had a 6 $\log_{10}(\mathrm{CFU/cm^2})$ biofilm, and by the end of 96 hours, there was no practical difference between biofilm growth on the coated catheters and control catheters.

Table 2 Antibiotic susceptibility assays

Method	Description	Pros	Cons	Ability to reflect CAUTI	Possible claim	Enumeration technique	References
Zone of inhibition	0.5-cm segments of catheter are embedded vertically in trypticase soy agar seeded on the surface with 10^8 CFU/mL of the test organism. The plates are incubated at 37 °C for 24 hours. The zones of inhibition recorded. Catheters transferred daily to fresh plates until no ZOI occurred	Rapid Cheap Simple Reproducible	Only dependent on chemistry's ability to diffuse into the agar Not useable for non-eluting surface modifications	Does not reflect environment seen during infection	Eluting chemistry Antibiotic susceptibility to clinically relevant pathogens	Zone of inhibition around the catheter segments, excluding the diameters of the catheters are measured	[61, 83–86]
Minimum inhibitory concentration (MIC)	Tubes containing increasing concentrations of antibiotic are inoculated with specific concentration of bacterium and incubated at 37 °C for 24 hours	Rapid Cheap Simple Reproducible	No surface tested	Does not test a surface's ability to prevent colonization or migration of bacteria Does not reflect environment seen during infection.	Antibiotic susceptibility to clinically relevant pathogens	MIC defined as lowest concentration of antibiotic at which there is no visible bacterial growth	[75, 87–89]
Liquid broth test	Surface-modified catheter segments are incubated at 37 °C for specified timespan in nutrient broth inoculated with specific concentration of bacterium. After incubation, segments are removed and washed. The segments are then placed in PBS and sonicated. Samples are serially diluted and plated for viable CFU counts	Surface is in constant contact with relevant media Simple Rapid Cheap Reproducible	Typically, not done for relevant time frame Static growth conditions	Simulates constant contact with urine Does not reflect environment seen during infection	Eluting chemistry Inhibition of growth of planktonic bacteria Prevents microbial adhesion under static conditions Antibiotic susceptibility to clinically relevant pathogens	Viable colony counts of attached bacteria Confocal laser scanning microscopy SEM Count of planktonic viable bacteria	[64, 79, 86, 89–91]

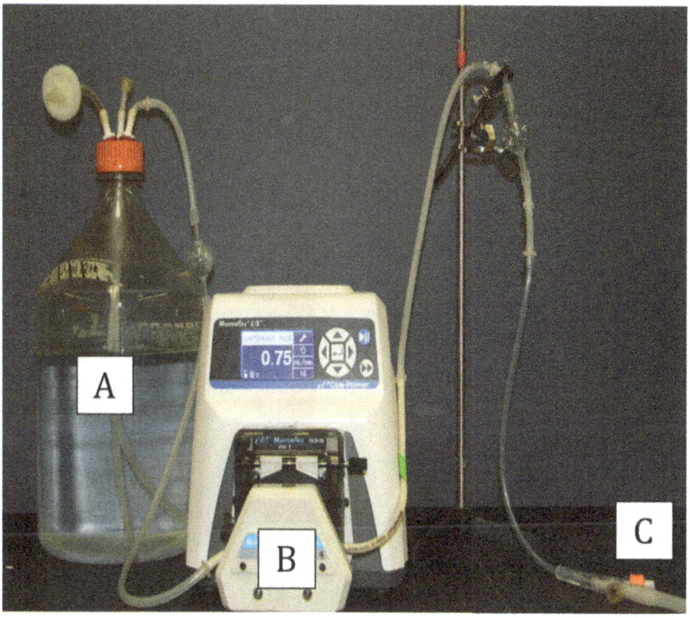

Fig. 4 Experimental set-up used to monitor biofilm formation on the intraluminal surface of urinary catheter. AUM (A) is pumped through tubing at 0.75 mL/min (B) through a 16 French (Fr) urinary catheter to sampling port (C)

Zone of Inhibition

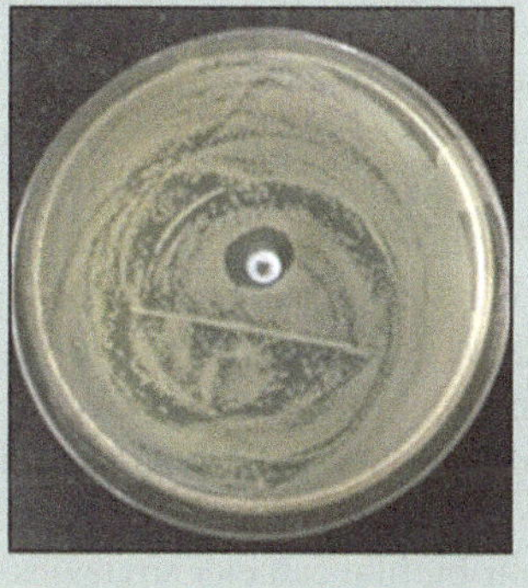

	Diameter(mm)
CHX1	14.7
CHX2	9.7
CHX3	9.7
Average ZOI	**11.3**

Liquid Broth Test

	Log (CFU/cm^2)
Control	5.9
Average CHX	2.3
Log reduction	3.6
% Reduction	**99.9**

Fig. 5 Left panel: ZOI pictured; ZOI was determined for 5 mm chlorhexidine-coated catheter segments. Right panel: Liquid broth test pictured; biofilm density was measured for control and chlorhexidine-coated catheters

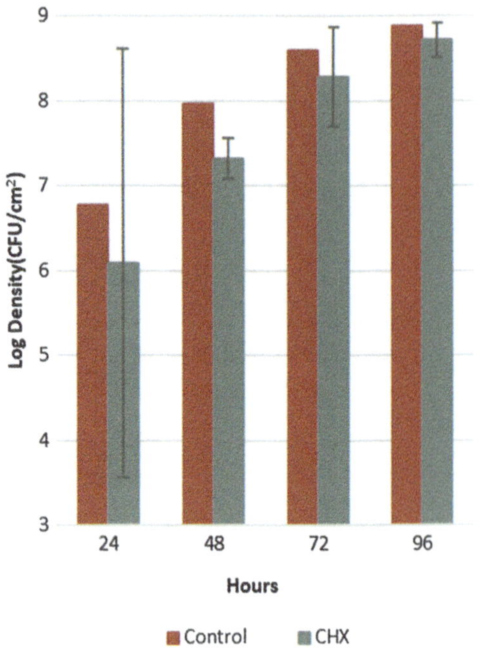

Fig. 6 Log densities of catheter samples from flow model. Error bars show the standard deviation of three coated catheters, and no error bars are reported on the control because of only one replicate tested

Based on the ZOI and liquid broth test results, the biofilm test results would not have necessarily been expected. The ZOI test demonstrated that chlorhexidine was able to diffuse into the surrounding environment and kill the viable microbes. The liquid broth test measured the bacterial adherence to the catheter surface. The test showed that chlorhexidine was able to inhibit adherence in static conditions for a short period of time, perhaps because the chlorhexidine killed the planktonic bacteria before they were able to attach. In the test where flow was included, chlorhexidine slowed down the biofilm growth at 24 hours by approximately 0.8 $\log_{10}$(CFU/cm^2). However, by 96 hours the concentration of chlorhexidine left on the catheter was insufficient to affect the thick biofilm that had developed on the intraluminal surface of the catheter. The results of this in vitro test comparison suggest that the size of zones produced, and log density of planktonic cells killed, may not accurately predict the effectiveness of antimicrobial urinary catheters against biofilm growth. The results also demonstrate the importance of including flow when considering performance claims for antimicrobial or anti-biofilm catheter strategies.

When thoughtfully engineered, an in vitro method can be used to better predict the clinical efficacy of a product. This section aims to examine the most frequently used laboratory methods in anti-biofilm urinary catheter testing and evaluate their usefulness based on [70]:

- *Relevance*: A method is said to be relevant to a real-world scenario if given the same inputs the laboratory outcome is predictive of the real-world outcome.
- *Repeatability*: Independent repeats of the same experiment in the same lab produce nearly the same response.

- *Responsiveness*: A method should be sensitive enough that it can detect important changes in parameters of interest.

In clinical terms, to reduce infection, a surface-modified catheter must delay the time it takes for bacteria to reach a concentration of 10^5 CFU/mL in the urine. When choosing a method to evaluate the efficacy of a new catheter, the outputs provided must be able to provide insight to how well it might be able to meet this criterion. Comprehensive lists of in vitro methods most commonly reported in the literature to model intraluminal and extraluminal infections are compared and contrasted in Tables 3 and 4, respectively. The intraluminal route of infection differs significantly from the extraluminal route. During intraluminal infection, the catheter is exposed to the flow of urine of the surface for extended periods, compared to extraluminal infections where there is a lack of flow but there is a complex relationship between the dynamic uroepithelium and catheter. Therefore, one model cannot accurately capture the complexity of both infections. Ideally, two models will be employed to evaluate the in vitro efficacy of the catheter to best predict in vivo performance.

Inherently in vitro methods are unable to completely replace in vivo methods because it is impossible to model the complex host-pathogen relationship. However, there are several factors which are typically neglected in an effort to simplify testing that may easily be included in laboratory tests to improve clinical correlation. Running tests for a relevant time frame is one critical parameter which is often dismissed in laboratory testing. Some methods, such as the liquid broth test, are typically only run for a period of hours when clinically an antimicrobial would be expected to delay infection over the course of several days. It is not reasonable to claim a surface modification will be effective when it was only exposed to a challenge for a few hours, especially in reference to intraluminal infections, where urine will flow over the surface for much longer. Simply increasing the experimental time or testing the device until failure will help make the data collected from an in vitro model more relevant.

Growth media is another notable factor that is often assumed to have little influence on the results. In vitro models are frequently conducted using a minimal, well-defined growth media (e.g., tryptic soy broth, lysogeny broth). It has been shown repeatably, though, that bacteria respond differently to varying growth medias, especially urine [54, 71, 72]. Due to its low pH and high osmolality, urine is naturally antimicrobial and creates an especially stressful environment for bacteria, ultimately resulting in altered metabolic pathways, virulence, and motility [29, 73, 74]. Not only does urine influence how bacteria grow and survive, it has also been shown to increase the MICs of several common antibiotics and decrease their overall efficacy [75]. It is therefore reasonable to assume that if urine is not used, results do not accurately predict how a catheter will perform in clinical settings.

Since urine has been identified as a necessary parameter for an in vitro model to be considered relevant, the decision becomes whether to use human urine or AUM. When considering how to make a method, the most repeatable AUM represents a good balance between reasonableness and relevance. The use of human urine does have some benefits; for example, it would contain natural constituents

Table 3 In vitro models to simulate intraluminal colonization by uropathogens

Method	Description	Pros	Cons	Ability to reflect CAUTI	Possible claim	Enumeration technique	References
Flow through	Inoculation of proximal end of catheter with test organism. Entire system is incubated at 37 °C. AUM is delivered at rate within laminar flow regimen. Effluent and catheter samples may be collected at multiple time points	Simulates in vivo environment Produces thick biofilm Can run for clinically relevant timespan Simple model set-up	Only evaluates ability to delay intraluminal colonization Uses large volumes of media	Simulates bacterial access to the intraluminal surface Mimics the spreading of an infection from one location	Efficacy of surface to delay bacterial adherence and migration on intraluminal surface of catheter in presence of urine flow	Viable colony counts from sampled urine and catheter segments Non-invasive optical biomass sensor SEM and TEM	[92, 93]
False bladder	False bladder (i.e., glass vessel) with AUM maintained at 37 °C. this allows a reservoir (~30 mL) of urine to collect in the bladder below the level of the catheter eyehole. As the volume of supplied urine increases, the overflow drains through the catheter to the collection bag. Residual urine in the model is inoculated directly. Samples collected from bladder, effluent flow and catheter segments	Simulates aspects of in vivo environment Able to relate bacteriuria to intraluminal growth	Does not simulate constant challenge of bacteria migrating up extraluminal surface Complex design decreases reproducibility across studies Uses large amount of media	Partially simulates "early" extraluminal inoculation	Ability for tip to "sterilize" bladder for time frame Ability to inhibit bacteria from gaining access to intraluminal surface Efficacy of surface to delay bacterial adherence and migration on intraluminal surface of catheter in presence of urine flow	Viable colony counts from sampled urine and catheter segments SEM and TEM	[46, 94–97]

Modified drip flow reactor (mDFR)	Device with four separate chambers with sealing lids. Catheter segments are connected to influent and effluent ports within the chambers. The device is connected to batch culture, which is pumped through at 1 mL/min for 2 hours, after that time sterile nutrient broth is pumped though lumen of catheters at 0.5 mL/min for specified timeframe. The catheters are then segmented, and biofilm is harvested for enumeration	Simulates constant low shear flow of urine over catheter surface Modified version of standardized method Produces thick biofilm	Uses large amount of media Shortened length of catheter could affect ability to reflect in vivo biofilm	Simulating bacterial access to the intraluminal surface	Efficacy of surface to delay bacterial adherence and migration on intraluminal surface of catheter in presence of urine flow	Viable colony count from catheter segments SEM	[62, 79, 98, 99]
Flow cell	Inoculum pumped across cell for 1 hour. Following inoculation period media is flowed through cell for given time period. Biofilm imaged using confocal microscopy	Simulates constant low shear flow of urine over catheter surface Allows real time microscopic investigation Uses little media	Expensive and expertise needed Shortened length of catheter could affect ability to reflect in vivo biofilm	Morphology of biofilm seen in flow cell does not typically represent that seen in infection [100]	Efficacy of surface to delay bacterial adherence and migration on intraluminal surface of catheter in presence of urine flow	Bioluminescence for monitoring biofilm accumulation	[79, 101–103]

(continued)

Table 3 (continued)

Method	Description	Pros	Cons	Ability to reflect CAUTI	Possible claim	Enumeration technique	References
Modified Robin device (MRD)	Ports sit linearly along rectangular channel. Plugs may be inserted into each port. Batch or continuous culture fluid is fed through the device, submerging the substrata. Biofilm density is sampled by removing plugs from the side of cell	Simulates constant low shear flow of urine over catheter surface	Simple disks of catheter material do not represent complex device	Simulating bacterial access to the intraluminal surface	Efficacy of surface to delay bacterial adherence and migration on intraluminal surface of catheter in presence of urine flow	Viable colony counts	[61, 104–106]

Table 4 Common in vitro methods used to model extraluminal migration of bacteria during CAUTI and evaluate the efficacy of surface-modified urinary catheters

Method	Description	Pros	Cons	Ability to reflect CAUTI	Possible claim	Enumeration technique	References
Anatomically correct	Growth media drips into catheterized bladder. Pools in bladder then drip around catheter down urethra (i.e., glass tube). Inoculum introduced around false meatus. Migration of bacteria tracked along the length of catheter and into the bladder	Able to run for extended timeframe Captures complexity of device	Complex design decreases reproducibility across studies Urethra-catheter interface not well represented with glass tube Uses large amounts of media	Attempted to simulate the physiological environment that the extraluminal is exposed to in vivo. Film of urine around extraluminal surface not seen clinically	Delay of bacteriuria and bacterial migration Along extraluminal surface	Viable colony counts from sampled urine and catheter segments	[46, 83]
False urethra	In a capped tube, agar is poured around a sealed segment of catheter creating an agar column. The top of this column has 3 mL of space which acts as the 'bladder' and at the bottom a portion of the catheter is exposed representing the 'meatus'. The meatus is inoculated with bacterium and bladder is sampled until bacterium is detected	Simplified model of anatomical structure Surrounds the catheter with nutrient media, creating pseudo-tissue environment Little media required	Significant issues with agar track decrease reproducibility across studies Bacteria find other routes to false bladder other than up catheter therefore rate of migration is not accurate	Attempted to simulate the physiological environment that the extraluminal is exposed to in vivo	Delay of bacteriuria and bacterial migration along extraluminal surface	Viable colony counts from sampled urine Roll plate method semi-quantitatively determines bacterial growth on surface	[84]

(continued)

Table 4 (continued)

Method	Description	Pros	Cons	Ability to reflect CAUTI	Possible claim	Enumeration technique	References
Modified motility assay	PDMS template with surface modification is inoculated with a uropathogen then placed face down on growth media, which has been poured on top of a microscope slide. System is incubated then imaged using phase contrast microscopy	Allows real time microscopic investigation Uses little media Test for non-eluting technologies	Expensive and expertise needed Cannot directly predict long term efficacy Does not contain final device in test as required by FDA[a]	Reflects the interstitial space between catheter and tissue	Delay of bacterial migration along extraluminal surface Inhibition of bacterial expansion on catheter surface	Low magnification phase contrast microscopy imaging and image analysis	[63, 107]
Minimum biofilm eradication concentration device (MBEC)	System consists of a conventional 96- well bottom plate and a top plate which has been modified so that there are 96 conical-shaped "pegs" with flattened tips protruding from the lid. Pegs can be dip coated with selected surface modification. During incubation, biofilms grow on the surface of the pegs and the nutrients for biofilm growth are replenished daily by aseptically transferring the lid to a new bottom plate containing fresh sterile medium	Commercially available device Uses little media Pegs can be coated in any material allowing for the screening of numerous treatments	Does not differentiate between matrix, living and dead cells attached to the surface Does not contain final device in test as required by FDA[b]	Does not reflect infection environment Does not reflect the catheter tissue interface seen in vivo	Measure efficacy of anti-biofilm technology to prevent biofilm adherence Measures total attached biomass	Bioluminescence for monitoring biofilm accumulation Crystal violet stain Optical density measurements Viable cell counts SEM	[79, 108–110]

Plate migration assay	1 cm channels are cut in nutrient agar plates. One side of channel is inoculated with uropathogen, 1 cm segment of catheter is placed in channel connecting the two sides. Plates are incubated and monitored for bacterial migration across catheter bridge	Simple rapid screening Little media needed Test for non-eluting technologies	Overly simplistic No relevant nutrients present	Does not reflect infection environment Does not reflect the catheter tissue interface seen in vivo	Delay of bacterial migration along extraluminal surface	Presence/absence	[111–113]

[a]https://www.fda.gov/MedicalDevices/ucm080884.htm. Accessed August 13, 2018
[b]https://www.fda.gov/MedicalDevices/ucm080884.htm. Accessed August 13, 2018

that are difficult to include in synthetic urine (i.e., hormones, iron chelators, and pyrophosphates). However, human urine is dynamic in composition, changing dramatically depending on age, gender, and health status [76–78]. A single commercial supplier offers over 15 different populations to choose from: normal, caffeine-free, nicotine-free, pregnancy first, second or third trimester, pre-menopausal women (on or off hormone modalities, including birth control), post-menopausal women (on or off hormone replacement therapies), drug-free (please specify drug), lipemic, fasted, or *race/ethnicity* (Black, Caucasian, Asian, Hispanic). In addition to wide chemical variation, volume limits are also a concern, with commercially available human urine costing up to $260/L.[3] Table 5 summarizes options for growth media available to researchers and benefits and drawbacks associated with each. More options do exist, as many have been developed depending on researchers' goals and specific applications. For example, Nowatzki et al. created their own AUM to better simulate the effects of salicylic acid release [79]. These recipes were developed with no intention for universal use, and most have chemical components well out of physiological range. A significant improvement would be the development of a validated AUM recipe which closely resembles "normal human urine" and supports a broad range of uropathogens. This would enable researchers in federal agencies, academia, and industry to compare the efficacy of surface-modified catheters across laboratories by eliminating the variability associated with the use of different growth media (Table 5).

The goal is to develop robust in vitro testing methods not to replace in vivo testing but to better predict how a new technology will perform in vivo. To accomplish this goal requires approaches in vitro methods that increase clinical relevance while maintaining the attributes of repeatability and responsiveness.

Conclusions and Path Forward

Urinary catheters are the leading cause of nosocomial infections not only nationwide but worldwide [80, 81]. Each infection costs patients anywhere from $1,000 to upwards of $10,000 depending on the severity of infection and patient population [82]. A large variety of biofilm methods are available to evaluate the efficacy of surface-modified urinary catheters. When employed thoughtfully, in vitro methods are a powerful tool, but one of the main issues for researchers is choosing a model which best represents the conditions seen in situ in a reasonable and repeatable way. As our knowledge of CAUTI pathogenesis continues to grow, the relevance of in vitro models will follow. The development of robust and relevant in vitro models will provide a pathway for new technology to reach patients.

[3] https://www.innov-research.com/product/normal-human-urine?c=1 Accessed June 29, 2018.

Table 5 Comparison of varying growth mediums used for in vitro models simulating CAUTIs

Growth medium	Description	Pro	Con	Ability to reflect	References
Human urine	Pooled urine, typically from 1 to 3 donors Normalized by adjusting dilution, concentration, and pH according to specific gravity and osmolality	Shown to support more abundant biofilm growth than artificial recipes [99] Support growth of broad range of uropathogens Contains natural components. Synthetic recipes cannot be incorporated	Expensive Variation between individuals is challenging for standardizing studies Large amounts of urine for models make use impractical	Human urine containing natural constitutes	[78, 88, 99]
Artificial urine media (AUM) within physiological ranges	Synthetic human urine with osmolality, pH and composition as close to physiological ranges as possible	Able to standardize across studies Concentrations of components within physiological range Support growth of broad range of uropathogens	Compositions depend on clinical urine composition reports researchers' reference Excludes some natural constituents of human urine (hormones, iron chelators)	Provides conditions similar to that of "normal human urine"	[88, 92, 99, 114–116]
Artificial urine media (AUM) for crystal aggregation	Elevated concentrations of specific solutes to encourage stone formation	Able to simulate infection for specific population/ disease	Certain chemical components out of physiological range Designed for specific infection, not planned for universal use Does not support growth of all uropathogens Potentially cytotoxic	Accurately represents environment in patients prone to developing kidney stones Not relevant for studies of general population	[54, 116–118]

(continued)

Table 5 (continued)

Growth medium	Description	Pro	Con	Ability to reflect	References
Nutrient broth	Minimal, well-defined growth media Broths commonly used include tryptic soy broth (TSB), lysogeny broth (LB) or Mueller-Hinton broth	Simple Reproducible Supports growth of broad range of uropathogens	Lacks all natural constitutes of human urine Metabolic activities of pathogens may be altered from those typically seen in infection MICs of antibiotics in nutrient broth are significantly lower than those observed in urine	Does not reflect environment seen during infection When used to determine bacterial susceptibility may not reflect the ability of bacteria in urine to resist the antibacterial action	[54, 64, 75, 88, 90]

Acknowledgments Darla M. Goeres, Ph.D., holds an Innovation in Regulatory Science Award from BWF.

References

1. Control, C.f.D., Urinary Tract Infection (Catheter-Associated Urinary Tract Infection [CAUTI] and Non-Catheter-Associated Urinary Tract Infection [UTI] and Other Urinary System Infection [USI] Events. 2018(Device-associated Module-UTI).
2. Lo, E., et al. (2014). Strategies to prevent catheter-associated urinary tract infections in acute care hospitals: 2014 update. *Infection Control & Hospital Epidemiology, 35*(Suppl 2), S32–S47.
3. Weiner, L. M., et al. (2016). Antimicrobial-resistant pathogens associated with healthcare-associated infections: Summary of data reported to the National Healthcare Safety Network at the Centers for Disease Control and Prevention, 2011–2014. *Infection Control and Hospital Epidemiology, 37*(11), 1288–1301.
4. Marzo, A., et al. (2018). An engineering approach towards a more discrete and efficient urinary drainage system. *Proceedings of the Institution of Mechanical Engineers. Part H*, 954411918790286.
5. Foley, F. E. B. (1937). A hemostatic bag catheter. *The Journal of Urology, 38*(1), 134–139.
6. Feneley, R. C., Hopley, I. B., & Wells, P. N. (2015). Urinary catheters: History, current status, adverse events and research agenda. *Journal of Medical Engineering & Technology, 39*(8), 459–470.
7. Hanafy, H. M., Saad, S. M., & Al-Ghorab, M. M. (1974). Ancient Egyptian medicine: Contribution to urology. *Urology, 4*(1), 114–120.
8. Miller, A., et al. (1958). Postoperative infection in urology. *Lancet, 2*(7047), 608–612.

9. Gillespie, W. A., et al. (1960). The diagnosis, epidemiology and control of urinary infection in urology and gynaecology. *Journal of Clinical Pathology, 13*, 187–194.
10. Garibaldi, R. A., et al. (1974). Factors predisposing to bacteriuria during indwelling urethral catheterization. *The New England Journal of Medicine, 291*(5), 215–219.
11. Garibaldi, R. A., et al. (1980). Meatal colonization and catheter-associated bacteriuria. *The New England Journal of Medicine, 303*(6), 316–318.
12. Costerton, J. W., Geesey, G. G., & Cheng, K. J. (1978). How bacteria stick. *Scientific American, 238*(1), 86–95.
13. Daifuku, R., & Stamm, W. E. (1986). Bacterial adherence to bladder uroepithelial cells in catheter-associated urinary tract infection. *The New England Journal of Medicine, 314*(19), 1208–1213.
14. Nickel, J. C., Gristina, A. G., & Costerton, J. W. (1985). Electron microscopic study of an infected Foley catheter. *Canadian Journal of Surgery, 28*(1), 50–51, 54.
15. Costerton, J. W., et al. (1987). Bacterial biofilms in nature and disease. *Annual Review of Microbiology, 41*, 435–464.
16. Donlan, R. M., & Issue, S. (2001). Biofilms and device-associated infections. *Emerging Infectious Diseases, 7*(2), 277–281.
17. Stoodley, P., et al. (2011). Orthopaedic biofilm infections. *Current Orthopaedic Practice, 22*(6), 558–563.
18. Busscher, H. J., et al. (2010). Biofilm formation on dental restorative and implant materials. *Journal of Dental Research, 89*(7), 657–665.
19. Hu, H., et al. (2016). Bacterial biofilm infection detected in breast implant-associated anaplastic large-cell lymphoma. *Plastic and Reconstructive Surgery, 137*(6), 1659–1669.
20. Meslemani, D., Yaremchuk, K., & Rontal, M. (2010). Presence of biofilm on adult tracheostomy tubes. *Ear, Nose, & Throat Journal, 89*(10), 496–504.
21. Food and Drug Administration, U. (1997). Guidance for the content of premarket notifications for conventional and antimicrobial Foley catheters.
22. Malpiedi, P. J. (2011). National and state healthcare-associated infection standardized infection ratio report.
23. Peasah, S. K., et al. (2013). Medicare non-payment of hospital-acquired infections: Infection rates three years post implementation. *Medicare & Medicaid Research Review, 3*(3).
24. Saint, S., et al. (2016). A program to prevent catheter-associated urinary tract infection in acute care. *The New England Journal of Medicine, 374*(22), 2111–2119.
25. Services, U.D.o.H.a.H. (2013). National action plan to prevent health care-associated infections: Road map to elimination, Executive Summary. p. 1–19.
26. Hooton, T. M., & Clinical, p. (2012). Uncomplicated urinary tract infection. *The New England Journal of Medicine, 366*(11), 1028–1037.
27. Ingersoll, M. A., & Albert, M. L. (2013). From infection to immunotherapy: Host immune responses to bacteria at the bladder mucosa. *Mucosal Immunology, 6*(6), 1041–1053.
28. Parsons, C. L., et al. (1990). Bladder surface glycosaminoglycans: An epithelial permeability barrier. *The Journal of Urology, 143*(1), 139–142.
29. Carlsson, S., et al. (2001). Effects of pH, nitrite, and ascorbic acid on nonenzymatic nitric oxide generation and bacterial growth in urine. *Nitric Oxide, 5*(6), 580–586.
30. Raffi, H. S., et al. (2005). Tamm-Horsfall protein acts as a general host-defense factor against bacterial cystitis. *American Journal of Nephrology, 25*(6), 570–578.
31. Hannan, T. J., et al. (2012). Host-pathogen checkpoints and population bottlenecks in persistent and intracellular uropathogenic Escherichia coli bladder infection. *FEMS Microbiology Reviews, 36*(3), 616–648.
32. Arinzon, Z., et al. (2012). Clinical presentation of urinary tract infection (UTI) differs with aging in women. *Archives of Gerontology and Geriatrics, 55*(1), 145–147.
33. de Lastours, V., & Foxman, B. (2014). Urinary tract infection in diabetes: Epidemiologic considerations. *Current Infectious Disease Reports, 16*(1), 389.

34. Lichtenberger, P., & Hooton, T. M. (2008). Complicated urinary tract infections. *Current Infectious Disease Reports, 10*(6), 499–504.
35. Levison, M. E., & Kaye, D. (2013). Treatment of complicated urinary tract infections with an emphasis on drug-resistant gram-negative uropathogens. *Current Infectious Disease Reports, 15*(2), 109–115.
36. Chenoweth, C. E., & Saint, S. (2011). Urinary tract infections. *Infectious Disease Clinics of North America, 25*(1), 103–115.
37. O'May, G. A., et al. (2009). Complicated urinary tract infections due to catheters., *3*, 123–165.
38. Haley, R. W., et al. (1981). Nosocomial infections in U.S. hospitals, 1975-1976: Estimated frequency by selected characteristics of patients. *The American Journal of Medicine, 70*(4), 947–959.
39. Maki, D. G., & Tambyah, P. A. (2001). Engineering out the risk for infection with urinary catheters. *Emerging Infectious Diseases, 7*(2), 342–347.
40. Jacobsen, S. M., et al. (2008). Complicated catheter-associated urinary tract infections due to Escherichia coli and Proteus mirabilis. *Clinical Microbiology Reviews, 21*(1), 26–59.
41. Stamm, W. E. (1991). Catheter associated urinary tract infections: Epidemiology, pathogenesis, and prevention. *The American Journal of Medicine, 91*(3B), 65S–71S.
42. Tambyah, P. A., Halvorson, K. T., & Maki, D. G. (1999). A prospective study of pathogenesis of catheter-associated urinary tract infections. *Mayo Clinic Proceedings, 74*(2), 131–136.
43. Saint, S., & Chenoweth, C. E. (2003). Biofilms and catheter-associated urinary tract infections. *Infectious Disease Clinics of North America, 17*(2), 411–432.
44. Nickel, J., Grant, S., & Costerton, J. (1985). Catheter-associated bacterium: An experimental study. *Urology, 26*(4), 369–375.
45. Warren, J. W. (2001). Catheter-associated urinary tract infections. *International Journal of Antimicrobial Agents, 17*(4), 299–303.
46. Barford, J. M., et al. (2008). A model of catheter-associated urinary tract infection inittiated by bacterial contamination of the catheter tip. *BJU International, 102*(1), 67–74.
47. Hidron, A. I., et al. (2008). NHSN annual update: Antimicrobial-resistant pathogens associated with healthcare-associated infections: Annual summary of data reported to the National Healthcare Safety Network at the Centers for Disease Control and Prevention, 2006–2007. *Infection Control and Hospital Epidemiology, 29*(11), 996–1011.
48. Mulvey, M. A. (2002). Adhesion and entry of uropathogenic Escherichia coli. *Cellular Microbiology, 4*(5), 257–271.
49. Struve, C., Bojer, M., & Krogfelt, K. A. (2008). Characterization of Klebsiella pneumoniae type 1 fimbriae by detection of phase variation during colonization and infection and impact on virulence. *Infection and Immunity, 76*(9), 4055–4065.
50. Stahlhut, S. G., et al. (2012). Biofilm formation of Klebsiella pneumoniae on urethral catheters requires either type 1 or type 3 fimbriae. *FEMS Immunology and Medical Microbiology, 65*(2), 350–359.
51. Fox-Moon, S.M. and M.E. Shirtliff, Urinary tract infections caused by Proteus mirabilis. 2015: p. 1389–1400.
52. Flores-Mireles, A. L., et al. (2015). Urinary tract infections: Epidemiology, mechanisms of infection and treatment options. *Nature Reviews. Microbiology, 13*(5), 269–284.
53. Nickel, J. C., et al. (1994). Bacterial biofilms: Influence on the pathogenesis, diagnosis and treatment of urinary tract infections. *The Journal of Antimicrobial Chemotherapy, 33*(Suppl A), 31–41.
54. Ipe, D. S., Horton, E., & Ulett, G. C. (2016). The basics of bacteriuria: Strategies of microbes for persistence in urine. *Frontiers in Cellular and Infection Microbiology, 6*, 14.
55. Justice, S. S., et al. (2008). Morphological plasticity as a bacterial survival strategy. *Nature Reviews. Microbiology, 6*(2), 162–168.
56. Stewart, P. S. (1996). Theoretical aspects of antibiotic diffusion into microbial biofilms. *Antimicrobial Agents and Chemotherapy, 40*(11), 2517–2522.
57. Hoyle, B. D., Alcantara, J., & Costerton, J. W. (1992). Pseudomonas aeruginosa biofilm as a diffusion barrier to piperacillin. *Antimicrobial Agents and Chemotherapy, 36*(9), 2054–2056.

58. Mobley, H. L., Island, M. D., & Hausinger, R. P. (1995). Molecular biology of microbial ureases. *Microbiological Reviews, 59*(3), 451–480.
59. Stickler, D., et al. (1993). Proteus mirabilis biofilms and the encrustation of urethral catheters. *Urological Research, 21*(6), 407–411.
60. Lee, C. R., et al. (2013). Strategies to minimize antibiotic resistance. *International Journal of Environmental Research and Public Health, 10*(9), 4274–4305.
61. Raad, I. (1995). Antibiotics and prevention of microbial colonization of catheters. *Antimicrobial Agents and Chemotherapy, 39*(11), 2397–2400.
62. Fu, W., et al. (2010). Bacteriophage cocktail for the prevention of biofilm formation by Pseudomonas aeruginosa on catheters in an in vitro model system. *Antimicrobial Agents and Chemotherapy, 54*(1), 397–404.
63. Gloag, E. S., et al. (2016). Micro-patterned surfaces that exploit stigmergy to inhibit biofilm expansion. *Frontiers in Microbiology, 7*, 2157.
64. Gabriel, M. M., et al. (1996). In vitro evaluation of the efficacy of a silver-coated catheter. *Current Microbiology, 33*(1), 1–5.
65. Bologna, R. A., et al. (1999). Hydrogel/silver ion-coated urinary catheter reduces nosocomial urinary tract infection rates in intensive care unit patients: A multicenter study. *Urology, 54*(6), 982–987.
66. Karchmer, T. B., et al. (2000). A randomized crossover study of silver-coated urinary catheters in hospitalized patients. *Archives of Internal Medicine, 160*(21), 3294–3298.
67. Srinivasan, A., et al. (2006). A prospective trial of a novel, silicone-based, silver-coated Foley catheter for the prevention of nosocomial urinary tract infections. *Infection Control and Hospital Epidemiology, 27*(1), 38–43.
68. Pickard, R., et al. (2012). Antimicrobial catheters for reduction of symptomatic urinary tract infection in adults requiring short-term catheterisation in hospital: A multicentre randomised controlled trial. *The Lancet, 380*(9857), 1927–1935.
69. Medline Industries, I., Silvertouch Foley Catheter: The Standard For Silver. 2009.
70. Parker, A. E., & Hamilton, M. A. (2011). *KSA-SM-10- Assessing Resemblance, Repeatability, and Reproducibility for quantitative methods*. Bozeman, MT: C.f.B.E.a.M.S. University.
71. Shrout, J. D., et al. (2006). The impact of quorum sensing and swarming motility on Pseudomonas aeruginosa biofilm formation is nutritionally conditional. *Molecular Microbiology, 62*(5), 1264–1277.
72. Palmer, K. L., Aye, L. M., & Whiteley, M. (2007). Nutritional cues control Pseudomonas aeruginosa multicellular behavior in cystic fibrosis sputum. *Journal of Bacteriology, 189*(22), 8079–8087.
73. Cotter, P. D., & Hill, C. (2003). Surviving the acid test: responses of gram-positive bacteria to low pH. *Microbiology and Molecular Biology Reviews, 67*(3), 429–453, table of contents.
74. Shepard, B. D., & Gilmore, M. S. (2002). Differential expression of virulence-related genes in Enterococcus faecalis in response to biological cues in serum and urine. *Infection and Immunity, 70*(8), 4344–4352.
75. Dalhoff, A., Stubbings, W., & Schubert, S. (2011). Comparative in vitro activities of the novel antibacterial finafloxacin against selected Gram-positive and Gram-negative bacteria tested in Mueller-Hinton broth and synthetic urine. *Antimicrobial Agents and Chemotherapy, 55*(4), 1814–1818.
76. Barr, D. B., et al. (2005). Urinary creatinine concentrations in the U.S. population: Implications for urinary biologic monitoring measurements. *Environmental Health Perspectives, 113*(2), 192–200.
77. Torffvit, O., & Agardh, C. D. (1993). Tubular secretion of Tamm-Horsfall protein is decreased in type 1 (insulin-dependent) diabetic patients with diabetic nephropathy. *Nephron, 65*(2), 227–231.
78. Bouatra, S., et al. (2013). The human urine metabolome. *PLoS One, 8*(9), e73076.
79. Nowatzki, P. J., et al. (2012). Salicylic acid-releasing polyurethane acrylate polymers as anti-biofilm urological catheter coatings. *Acta Biomaterialia, 8*(5), 1869–1880.

80. Wang, J., et al. (2017). Burden of healthcare-associated infections in China: Results of the 2015 point prevalence survey in dong Guan City. *The Journal of Hospital Infection, 96*(2), 132–138.
81. Zarb, P., et al. (2012). The European Centre for Disease Prevention and Control (ECDC) pilot point prevalence survey of healthcare-associated infections and antimicrobial use. *Euro Surveillance, 17*(46).
82. Hollenbeak, C. S., & Schilling, A. L. (2018). The attributable cost of catheter-associated urinary tract infections in the United States: A systematic review. *American Journal of Infection Control, 46*(7), 751–757.
83. Darouiche, R. O. (1997). In vitro efficacy of antimicrobial-coated bladder catheters in inhibiting bacterial migration along catheter surface. Concise Communications. *The Journal of Infectious Diseases, 176*, 1109–1112.
84. Gaonkar, T. A., et al. (2003). Evaluation of the antimicrobial efficacy of urinary catheters impregnated with antiseptics in an in vitro urinary tract model. *Infection Control and Hospital Epidemiology, 24*(7), 506–513.
85. Sherertz, R. J., Forman, D. M., & Solomon, D. D. (1989). Efficacy of dicloxacillin-coated polyurethane catheters in preventing subcutaneous Staphylococcus aureus infection in mice. *Antimicrobial Agents and Chemotherapy, 33*(8), 1174–1178.
86. Darouiche, R. O., et al. (2008). Efficacy of combination of chlorhexidine and protamine sulphate against device-associated pathogens. *The Journal of Antimicrobial Chemotherapy, 61*(3), 651–657.
87. Gomes, L. C., et al. (2015). Escherichia coli adhesion, biofilm development and antibiotic susceptibility on biomedical materials. *Journal of Biomedical Materials Research. Part A, 103*(4), 1414–1423.
88. Minuth, J. N., Musher, D. M., & Thorsteinsson, S. B. (1976). Inhibition of the antibacterial activity of gentamicin by urine. *The Journal of Infectious Diseases, 133*(1), 14–21.
89. Dave, R. N., Joshi, H. M., & Venugopalan, V. P. (2011). Novel biocatalytic polymer-based antimicrobial coatings as potential ureteral biomaterial: Preparation and in vitro performance evaluation. *Antimicrobial Agents and Chemotherapy, 55*(2), 845–853.
90. Hachem, R., et al. (2009). Novel antiseptic urinary catheters for prevention of urinary tract infections: Correlation of in vivo and in vitro test results. *Antimicrobial Agents and Chemotherapy, 53*(12), 5145–5149.
91. Roe, D., et al. (2008). Antimicrobial surface functionalization of plastic catheters by silver nanoparticles. *The Journal of Antimicrobial Chemotherapy, 61*(4), 869–876.
92. Nickel, J. C., Downey, J., & Costerton, J. W. (1992). Movement of pseudomonas aeruginosa along catheter surfaces. A mechanism in pathogenesis of catheter-associated infection. *Urology, 39*(1), 93–98.
93. Dohnt, K., et al. (2011). An in vitro urinary tract catheter system to investigate biofilm development in catheter-associated urinary tract infections. *Journal of Microbiological Methods, 87*(3), 302–308.
94. Stickler, D. J., Morris, N. S., & Winters, C. (1999). Simple physical model to study formation and physiology of biofilms on urethral catheters. *Methods in Enzymology, 310*, 494–501.
95. Tenke, P., et al. (2004). Bacterial biofilm formation on urologic devices and heparin coating as preventive strategy. *International Journal of Antimicrobial Agents, 23 Suppl 1*, S67–S74.
96. Chua, R. Y. R., et al. (2017). An in vitro urinary catheterization model that approximates clinical conditions for evaluation of innovations to prevent catheter associated urinary tract infections. *The Journal of Hospital Infection, 97*, 66.
97. Morris, N. S., Stickler, D. J., & Winters, C. (1997). Which indwelling urethral catheters resist encrustation by Proteus mirabilis biofilms? *British Journal of Urology, 80*(1), 58–63.
98. Curtin, J. J., & Donlan, R. M. (2006). Using bacteriophages to reduce formation of catheter-associated biofilms by Staphylococcus epidermidis. *Antimicrobial Agents and Chemotherapy, 50*(4), 1268–1275.

99. Lehman, S. M., & Donlan, R. M. (2015). Bacteriophage-mediated control of a two-species biofilm formed by microorganisms causing catheter-associated urinary tract infections in an in vitro urinary catheter model. *Antimicrobial Agents and Chemotherapy, 59*(2), 1127–1137.
100. Roberts, A. E., et al. (2015). The limitations of in vitro experimentation in understanding biofilms and chronic infection. *Journal of Molecular Biology, 427*(23), 3646–3661.
101. Purevdorj, B., Costerton, J. W., & Stoodley, P. (2002). Influence of hydrodynamics and cell signaling on the structure and behavior of Pseudomonas aeruginosa biofilms. *Applied and Environmental Microbiology, 68*(9), 4457–4464.
102. Heydorn, A., et al. (2000). Experimental reproducibility in flow-chamber biofilms. *Microbiology, 146*(Pt 10), 2409–2415.
103. Tolker-Nielsen, T., & Sternberg, C. (2011). Growing and analyzing biofilms in flow chambers. *Current Protocols in Microbiology*. Chapter 1: p. Unit 1B 2.
104. Nickel, J. C., et al. (1985). Antibiotic resistance of Pseudomonas aeruginosa colonizing a urinary catheter in vitro. *European Journal of Clinical Microbiology, 4*(2), 213–218.
105. Maeyama, R., et al. (2005). Novel bactericidal surface: Catechin-loaded surface-erodible polymer prevents biofilm formation. *Journal of Biomedical Materials Research. Part A, 75*(1), 146–155.
106. McCoy, W. F., et al. (1981). Observations of fouling biofilm formation. *Canadian Journal of Microbiology, 27*(9), 910–917.
107. Turnbull, L., & Whitchurch, C. B. (2014). Motility assay: Twitching motility. *Methods in Molecular Biology, 1149*, 73–86.
108. Irwin, N. J., McCoy, C. P., & Carson, L. (2013). Effect of pH on the in vitro susceptibility of planktonic and biofilm-grown Proteus mirabilis to the quinolone antimicrobials. *Journal of Applied Microbiology, 115*(2), 382–389.
109. Ceri, H., et al. (1999). The Calgary Biofilm Device: New technology for rapid determination of antibiotic susceptibilities of bacterial biofilms. *Journal of Clinical Microbiology, 37*(6), 1771–1776.
110. Meije, Y., et al. (2014). Daptomycin is effective as antibiotic-lock therapy in a model of Staphylococcus aureus catheter-related infection. *The Journal of Infection, 68*(6), 548–552.
111. Stickler, D., & Hughes, G. (1999). Ability of Proteus mirabilis to swarm over urethral catheters. *European Journal of Clinical Microbiology & Infectious Diseases, 18*(3), 206–208.
112. Sabbuba, N. (2002). The migration of Proteus Mirabilis and other urinary tract pathogens over Foley catheters. *BJU International, 89*, 55–60.
113. Reddy, S. T., et al. (2011). Micropatterned surfaces for reducing the risk of catheter-associated urinary tract infection: An in vitro study on the effect of sharklet micropatterned surfaces to inhibit bacterial colonization and migration of uropathogenic Escherichia coli. *Journal of Endourology, 25*(9), 1547–1552.
114. Burns, J. R., & Finlayson, B. (1980). A proposal for a standard reference artificial urine in in vitro urolithiasis experiments. *Investigative Urology, 18*(2), 167–169.
115. Brooks, T. (1997). A simple artificial urine for the growth of urinary pathogens. *Letters in Applied Microbiology, 24*, 203–206.
116. Chutipongtanate, S., & Thongboonkerd, V. (2010). Systematic comparisons of artificial urine formulas for in vitro cellular study. *Analytical Biochemistry, 402*(1), 110–112.
117. Grases, F., & Llobera, A. (1998). Experimental model to study sedimentary kidney stones. *Micron, 29*(2–3), 105–111.
118. Christmas, K. G., et al. (2002). Aggregation and dispersion characteristics of calcium oxalate monohydrate: Effect of urinary species. *Journal of Colloid and Interface Science, 256*(1), 168–174.

Biofilms and Wound Infection Research in the US Military

Kevin S. Akers, Joseph C. Wenke, and Clinton K. Murray

Abstract Recent US military conflicts have involved severe extremity injuries frequently requiring implantation of orthopedic stabilizing devices. Simultaneously, bacterial wound contamination, including by multidrug-resistant organisms, has presented a significant clinical challenge due to reduced antimicrobial treatment options, with an unclear but likely contribution from biofilm formation on implanted devices. In this chapter, we detail investigations conducted by the US military medical research community into wound infections occurring in casualties from conflicts in Iraq and Afghanistan.

Keywords Military · Conflicts · Wound · Biofilm · Antimicrobial · Research

Introduction

US military combat operations in Iraq (Operation Iraqi Freedom, OIF) and Afghanistan (Operation Enduring Freedom, OEF) frequently involved a high number of extremity injuries. This is likely multifactorial, influenced by the use of body armor protecting the trunk but not the extremities, tourniquets permitting survival to surgical care by controlling potentially lethal extremity hemorrhage, and enemy tactics involving the use of improvised explosive devices (IEDs). In Afghanistan, IEDs were frequently buried underground and detonated when military personnel

K. S. Akers (✉)
United States Army Institute of Surgical Research, Fort Sam Houston, TX, USA

Uniformed Services University of the Health Sciences, Bethesda, MD, USA
e-mail: Kevin.s.akers.mil@mail.mil

J. C. Wenke
United States Army Institute of Surgical Research, Fort Sam Houston, TX, USA

C. K. Murray
Uniformed Services University of the Health Sciences, Bethesda, MD, USA

1st Area Medical Laboratory, Aberdeen Proving Ground, Aberdeen, MD, USA

D. L. Williams (ed.), *Targeting Biofilms in Translational Research, Device Development, and Industrial Sectors*,
https://doi.org/10.1007/978-3-030-30667-0_4

on foot (known as a dismounted patrol) were in the vicinity or overtop of the ordnance, resulting in an injury pattern which came to be known as a "dismounted complex blast injury." Such injuries generally involved the lower extremities and possibly the groin and were frequently characterized by extensive soft tissue damage, vascular injury, and complex fractures requiring hardware stabilization and definitive fixation to achieve fracture healing. A further complication is that both trauma and transfusion are significantly immunomodulatory, with little known about the impact on clinical outcome or susceptibility to infection [1].

Contemporaneously, bacterial pathogens with broad antimicrobial resistance became established in US military facilities due to a convergence of factors: (1) a growing global pandemic of bacteria harboring various mechanisms of resistance to antimicrobials, known as multidrug-resistant organisms (MDROs); (2) care provided to local civilians, representing a source of MDROs entering deployed military medical facilities [2]; and (3) challenges with adherence to infection control measures in these facilities, particularly during mass casualty events and intercontinental air transport of combat casualties [3]. The resulting clinical circumstance of severe complex traumatic extremity wounds requiring indwelling hardware with possible contamination by MDRO organisms sets the stage for a multiyear outbreak of combat-related extremity wound infection in the US military healthcare system [4]. Although clinical confirmation was, and remains, limited by a lack of available diagnostic tests, bacterial biofilms colonizing implanted orthopedic fixation devices may further complicate the care of these patients by leading to persistent or relapsing infections [5].

Clinical Impact of Extremity Injury in US Military Combat Casualties

The impact of combat-related extremity injuries on US military personnel from OIF/OEF was documented in a series of studies performed at the US Army Institute of Surgical Research and several military treatment facilities in the United States. Extremity injuries were highly prevalent, with approximately 82% of combat casualties having at least one extremity injury [6]. Overall, 24–27% of extremity injuries developed infection, with 17% of those experiencing recurrence [7, 8]. The treatment of extremity wound infection typically involves multiple debridement surgeries followed by antibiotic therapy for weeks to months. Wound dehiscence was a major problem in the treatment of combat wounds, with wound debridement for infection being the most common procedure in battlefield injured service members. Approximately 12% of casualties required hospital readmission for wound infection debridement [9]. This in turn was a leading cost driver of medical care for US military combat casualties.

Not unexpectedly, the wound infection problem has been a major detriment to combat casualties with lower extremity orthopedic injuries, leading to increased amputation rates and decreased return-to-duty rates. Among 115 military casualties with type III open tibial fractures sustained between 2003 and 2007, the return-to-duty rate was just 18% if tibial fracture was the only injury, decreasing to 12.5% if an amputation was performed [10]. Illustrating the impact of infection, 92% of those with osteomyelitis were medically retired from military service. Sixty-nine percent were unfit for service due to a tibial fracture, and those with amputations received a higher Veteran's Affairs (VA) disability rating. A second study of military casualties receiving amputations between 2001 and 2006 found return-to-duty rates of 18–22% for tibial or femoral amputations, and 17% for humeral or radial amputations [11]. In a cohort of 115 soldiers with a type III open tibial fracture, 40% suffered an infectious complication and 94/115 (81.7%) were unable to return to duty and were medically retired [12]. In a study of late amputations after combat-related type III open tibial fractures, patients receiving late amputations had the highest rates of deep soft tissue infection (72.7%) and osteomyelitis (54.5%). Those who had late amputations had higher rates of grafting with autologous bone or bone morphogenic protein (BMP) combined than those with successful limb salvage (27.3% vs 4.8%, $p < 0.01$) [13].

Clinical Impact of Trauma-Related Infection in US Military Combat Casualties

The Trauma Infectious Disease Outcomes Study (TIDOS) [14] prospectively collected injury and infection data from combat casualties from Iraq and Afghanistan between 2009 and 2015. While analysis of these data is ongoing, a preliminary report has summarized the first 3 years of data (2009–2012) [15]. Over this period, 1807 combat casualties were included in the study, with 34% having an infection. Of those with infections, 57% had more than one infection. Independent risk factors for infection included amputation, blood transfusion within 24 hours of injury, ICU admission at Landstuhl Regional Medical Center (LRMC), severe or life-threatening Injury Severity Score, and mechanical ventilation.

Johnson and colleagues at the Brooke Army Medical Center Infectious Disease Service reported on infectious outcomes of patients with Gustilo-Anderson type III open tibial fractures (11 type IIIa and 24 type IIIb) in US military personnel occurring in Iraq and Afghanistan between 2003 and 2006 [16]. Seventy-seven percent were associated with an explosive device as the mechanism of injury. Casualties were managed initially with perioperative debridement including use of Gram-positive coverage with vancomycin or cefazolin. They were transported to LRMC in Germany, where a single debridement surgery was typically performed prior to transport back to the United States for definitive medical care, on average 7.4 days after injury. Twenty-seven (77%) had positive initial cultures at the time of admission to a US hospital. The most commonly recovered organisms in culture were *A. baumannii-calcoaceticus* complex, *Enterobacter* spp., and *P. aeruginosa.*

Polymicrobial infection was documented in 10 cases (37%). Most patients (89%) were treated for osteomyelitis, with only three treated for deep wound infection. Thirteen cases (37%) were complicated by a recurrence of deep wound infection involving the fracture site, of which 11 had previous infections. Most notably, almost all the initial infections involved Gram-negative organisms, whereas all of the recurrent infections involved *S. aureus*. Additionally, culture-positive infection was a contributing factor to four of five amputations. The fifth amputation was associated with tibial non-union without culture-proven evidence of infection [16].

Following this report, Yun and colleagues from the same institution examined cases of osteomyelitis (not restricted to tibial fractures) in military personnel injured in Iraq and Afghanistan over approximately the same time period [17]. They analyzed 101 patients who experienced 103 initial and 36 subsequent hospitalizations, with 94 involving lower extremities, 43 involving upper extremities, and 2 involving the axial skeleton. Recurrent infection occurred in 19 patients (18%). While Injury Severity Score was not found to be different between patients with non-recurrent and recurrent infections, Gram-negative organisms were more likely to occur in the initial infection, particularly involving *A. baumannii-calcoaceticus* complex, *Klebsiella pneumoniae*, and *P. aeruginosa*. Similar to Johnson's report [16], recurrences were associated with recovery of Gram-positive organisms, particularly *S. aureus* and coagulase-negative staphylococci, and less likely to be polymicrobial. Suggesting a contribution from biofilms in recurrent infections, the presence of an internal fixation device was the only factor found to be associated with recurrence of infection in a univariate analysis. Ninety percent of these patients received antimicrobial therapy for greater than 4 weeks, and 78% were treated longer than 6 weeks. An important finding from this analysis was that among patients with recurrence involving methicillin-resistant *S. aureus* (MRSA) in whom it was not present in the initial infection, 67% had received vancomycin for greater than 2 weeks. This finding highlights the important role of antimicrobial stewardship in the treatment of such patients and argues against the presumptive or prophylactic use of vancomycin without an appropriate microbiological indication [17].

Data from the Department of Defense Trauma Registry (previously called the Joint Theater Trauma Registry) [18] enabled a case-control study to determine risk factors for osteomyelitis among combat casualties with open tibial fractures occurring between 2003 and 2009 [19]. One-hundred thirty cases (patients with infected open tibial fractures) were compared to 85 controls (patients with open tibial fractures without infection at that site). The Gustilo-Anderson fracture classification was modified to include trans-tibial amputation (TTA) as the most severe grade. Excluding TTA, cases required longer time to achieve radiographic union (median, 210 versus 165 days). Risk factors for osteomyelitis included blast mechanism of injury, utilization of antibiotic beads, Gustilo-Anderson grade greater than IIIb, and presence of foreign bodies at the fracture site. Notably, TTA carried the highest risk of infection. A separate analysis utilizing the Orthopedic Trauma Association Open Fracture Classification system demonstrated a spectrum of increasing infection risk associated from muscle loss to muscle death [19].

The TIDOS data support a very important and unique longitudinal collaborative effort with the Department of Veterans' Affairs (VA) hospital system [20]. The purpose is to document long-term infection risk and outcomes among combat casualties after being discharged from the military after which time their medical care often transitions to the VA system. Although work is still ongoing to fully characterize the cohort, an initial publication documents the findings from the initial 337 for whom complete data have been abstracted and analyzed. One-hundred eleven (33%) had at least one infection related to their traumatic injury during hospitalization in the US Department of Defense medical system (totaling 244 unique infections). The most common were skin/soft tissue infection (SSTI, 43%) and osteomyelitis (14%). After the initial hospitalization, 127 patients (38%) developed 239 new infections related to their traumatic injury. Twenty-nine percent of these infections occurred after leaving military service. Independent risk factors for reduced time to infection following initial hospitalization were more severe injuries (Injury Severity Score > 10) and having an infection during the initial hospitalization. Notable findings within the VA phase of care were a "second peak" of SSTI and osteomyelitis 6–12 months after the initial infection and more frequent diagnosis of urinary tract infections, potentially reflecting perineal trauma from dismounted complex blast injury [20].

The Role of Biofilms in Extremity Wound Infection

Many combat-related extremity injuries are complex, involving extensive bone and soft tissue damage and requiring the presence of artificial materials to maintain the spatial orientation and anatomic alignment of bone fragments necessary to achieve an acceptable result of healing [21]. An unintended consequence of hardware required to stabilize mangled extremities is that they may serve as a substrate for bacterial attachment and the development of biofilms. Biofilms can develop when bacteria expand horizontally and vertically on a solid surface, forming a sessile, multicellular colony which secretes a matrix of protein, polysaccharide, and extracellular DNA that impedes the penetration of antimicrobials [22]. Bacteria in the lower strata of the colony cease replication and thus become tolerant to currently approved antimicrobials and create persister cells. In addition, the degree of soft tissue damage and vascular disruption in these extensive wounds likely limits antimicrobial penetration to site of infection. As a result, systemic treatment regimens with currently approved antimicrobial agents cannot predictably eradicate biofilms [23], which therefore pose a risk for infectious relapse in the setting of devices that must be retained for healing. Thus, adequate surgical debridement is required to mitigate these infections. A limitation to this approach is that there is no objective way for surgeons to judge when debridement has been sufficient and final wound closure can be performed with a minimal risk of infectious relapse. Complicating and adding to this, the soft tissue injuries are so severe that the wounds cannot be closed immediately, often being debrided in different operating rooms while the patient is being evacuated from the Middle East to Germany, and eventually arriving at military hospitals in the United States. Unfortunately, the patient records are often incomplete and may not accompany the Wounded Warrior, leading to a common practice of writing last time of debridement and treatment of wound on the exterior of dressings with a marker. Thus, a biomarker that could be used to accurately and objectively predict successful wound closure would be highly advantageous.

Studies by US military-affiliated investigators have included both natural history and basic science approaches. Investigators at the Walter Reed National Naval Medical Center and nearby Uniformed Services University of the Health Sciences in Bethesda, Maryland, evaluated the wound effluent captured from negative-pressure wound therapy for biomarkers useful to predict wound dehiscence. Forsberg and colleagues found an association between non-healing wounds and decreased effluent levels of RANTES and IL-13 [24]. Hawksworth and colleagues then examined other inflammatory biomarkers in serum and effluent, finding that increased serum IL-6, IL-8, and MIP-1α predicted wound healing. In the effluent, increased IL-6 and decreased IL-2 and IP-10 predicted healing [25]. Utz and colleagues examined metalloproteinases (MMP), finding that increased serum MMP-2 and MMP-7, and decreased effluent MMP-3, predicted impaired healing [26]. Brown and colleagues examined the microbial burden in combat wounds, defining a threshold of critical bacterial colonization of the wound as $>10^5$ CFU/g above which wound dehiscence is more likely to occur. Dehiscence was predicted by increased serum IL-6, -8, -10, MIP-1α, MMP-7, and MMP-13, as well as increased IL-6, IL-8, and MIP-1α in wound effluent [27]. Evans and colleagues examined the relationship of critical bacterial colonization (here defined as $>10^5$ CFU/cm^3) to heterotopic ossification (HO), finding that increased IL-6, IL-10, and MCP-1 in serum, and increased MIP-1α and decreased IP-10 in effluent, predicted development of HO [28].

Assessments of the clinical impact of biofilm are limited. At the US Army Institute of Surgical Research, Sanchez et al. investigated the biofilm production by 205 clinical strains of bacteria obtained from various solid and liquid source tissues of 150 patients with relapsing infections [29]. Among these isolates from wound, bone, respiratory, urinary tract, and blood isolates, later-recovered isolates were identical clones (by pulsed-field gel electrophoresis) of initially recovered strains. Biofilm formation, characterized by crystal violet staining of 48-hour growth in vitro, was heterogeneous, with increased biofilm production among isolates recovered from non-liquid sources.

Using prospectively collected data from the Trauma Infectious Disease Outcomes Study (TIDOS) [14], we examined risk factors including for the persistence of clinical wound infections meeting CDC/NHSN criteria for skin and soft tissue infection [5]. This study defined "persistence" as recovery of clonally identical isolates at least 14 days apart. Thirty-five persistently infected wounds from 25 patients were compared to 69 wounds from 60 patients with a single episode of wound infection. We identified biofilm formation to be a univariate risk factor (OR 29.49, 95% CI 6.24–∞) but lacked sufficient clinical data to associate this finding with implanted medical devices. Further, univariate risk factors also included MDRO phenotype, packed red blood cell transfusion within the first 24 hours after injury, number of operating room visits prior to and on the date of infection diagnosis, anatomic location of infection, and polymicrobial infection. Independence of these risk factors for prediction of infection persistence unfortunately could not be established, as the small sample size prevented convergence of a multivariate logistic regression model. *Acinetobacter baumannii* was the only species for which higher biofilm formation was statistically associated with wound infection persistence. This study suggests a possible linkage between biofilm formation as a phenotypic trait and persistent wound infection.

Heitcamp and colleagues utilized the TIDOS database to explore the role of *Enterococcus* spp. in clinical infection, as it was among the most frequent clinically isolated organisms in this data set [30]. Using a case-comparator study design, Heitcamp and colleagues compared 155 cases having *Enterococcus* spp. recovered within 3 days of wound infection diagnosis to 237 comparators for whom other organisms (but not *Enterococcus* spp.) were recovered in this time frame. *E. faecium* was the most common species (65.7%) followed by *E. faecalis* (12.5%), with most isolates arising in the setting of multiple and polymicrobial infections. Case patients were more likely to have a higher injury severity score and an increased rate of ICU admission and have received more transfused units of packed red blood cells and/or whole blood within 24 hours of injury, more operating room visits, and a longer length of hospitalization. Although biofilms were not specifically addressed in this manuscript, biofilm characterization of TIDOS *Enterococcus* isolates performed at the US Army Institute of Surgical Research showed almost universally low biofilm formation (unpublished data). An ensuing collegial debate about this finding among members of the TIDOS investigative team highlighted the question of what is the most appropriate manner for growth and characterization of biofilm-forming potential of clinical isolates. We previously explored this issue in the laboratory using clinical and reference strains of *S. aureus*, characterizing biofilm growth without human plasma, or with various concentrations of plasma either coating the growth plate, or added to the liquid growth medium. We found *S. aureus* to be optimally stimulated by addition of 10% human plasma to growth medium, increasing the crystal violet uptake signal by between 2.5- and nearly ten-fold depending on the strain. This was accompanied by dramatic fold changes in the expression of genes regulating matrix adhesion molecules (MSCRAMMs) between 30 and 120 minutes, with some increasing and others decreasing [31]. As growth conditions reported in biofilm literature vary widely (which we have previously reviewed [32]), and the biofilm production of some organisms is known to be significantly impacted by host factors likely to be present in wounds [31], a consensus method is currently lacking. Harmonization of biofilm testing conditions for clinical strains which accurately recapitulates in vivo biofilm-productive behavior in clinical infections would be a significant advance for the field.

Strategies for Biofilm Mitigation

Research efforts within the US military medical research enterprise have targeted potential approaches by which to mitigate biofilm formation in wounds. At the US Army Institute of Surgical Research, Wenke and colleagues established the rat femur segmental bone defect model to conduct research on various aspects of orthopedic trauma [33]. This model attempts to recapitulate the time course of care provided for human long-bone fractures, with delayed presentation to treatment, followed by irrigation, debridement, and long-bone stabilization by fixation with implanted materials. The carefully controlled addition of a bacterial inoculum to establish what is likely a biofilm-mediated orthopedic device infection makes this model a useful platform by which to explore and evaluate optimal approaches to mitigate biofilm formation. Favorable results can then be "scaled up" into larger animal models. Studies to date (some using goats) have characterized optimal approaches to fluid lavage of contaminated wounds [34], various antiseptic additives [35] and chlorhexidine [33], and Dakin's solution (buffered hypochlorite) as well as proprietary substances. A unifying theme which has emerged from this work is that local tissue damage within the wound, whether chemical or mechanical/physical, promotes bacterial growth in spite of initial reductions in bacterial burden. Therefore, an ideal substance for topically applied prophylaxis or treatment of wound infections would be one that is rapidly lethal to pathogens but innocuous to host mammalian tissues. Whereas conventional toxicity testing, driven by regulatory requirements for the cosmetic and topical medication industries, relies on skin cells such as keratinocytes and fibroblasts, we have noted that cell types relevant to deeper wound and bone tissues (such as myocytes, osteoblasts, etc.) seem to be more vulnerable to toxic insult [36–38]. Importantly, the regulatory approach for topical antiseptics has relied on toxicity testing using intact skin as opposed to deeper tissues which become exposed and contaminated in severe blast injuries or high-velocity penetrating trauma.

In addition to topical antiseptics, we have examined several conventional systemic antibiotics for potential repurposing as topical antimicrobials applied to contaminated wounds. This concept evolved from clinical reports indicating a decrease in infection rates associated with human spine surgery when vancomycin powder was topically applied as a prophylactic within clean surgical wounds [39–41]. Vancomycin powder and polymethyl methacrylate (PMMA) beads impregnated with 10% vancomycin (wt/wt) were examined in the rat femoral segmental defect model inoculated with *Staphylococcus aureus* UAMS-1. Animals received debridement and vancomycin powder or vancomycin-impregnated PMMA bead placement either 6 or 24 hours after contamination [42]. Significant reductions in bacterial burden were observed when either treatment was applied 6 hours after inoculation but failed to prevent wound infection when applied 24 hours after inoculation. This is postulated to represent the effect of biofilm maturation on the implanted orthopedic stabilizing device over this time period. Thus, the time elapsed from a contaminating event until debridement may constitute an important factor influencing the outcome of wound infection. Early irrigation was found beneficial in a goat contaminated wound model, perhaps denying an opportunity for biofilms to become established [43]. In humans, longer times to debridement of open tibial/fibular fractures (a condition known to be high risk for infection) were correlated with increased rates of clinical infection [44]. Another unifying theme has been that if there is a substantial time delay from time of contamination to treatment, an approach that addresses biofilms is needed.

Rifampin has a unique role among antimicrobial agents currently approved for use in humans, as it is considered the most suitable agent for the treatment of bacterial biofilms. Rifampin is less soluble in water than most other human-use antibiotics and has been demonstrated in the laboratory to have an increased capacity for physical penetration into biofilms [45]. Despite relatively limited human clinical data, the use of systemic rifampin in combination with other antimicrobials is recommended for prosthetic valve endocarditis and osteoarticular infections involving prosthetic devices (thought to involve biofilms) in guidelines for the treatment of methicillin-resistant *S. aureus* (MRSA) infections by the Infectious Disease Society of America [46].

We explored in our laboratory the potential utility of topically applied rifampin delivered in PMMA beads [47]. A common practice in orthopedic surgery, PMMA is polymerized from two components in an exothermic chemical reaction and mixed at the time of use, curing to a hard, rigid material within 15 minutes. Heat-stable antibiotic powders such as vancomycin, gentamicin, or tobramycin can be added at the surgeon's discretion. Using the rat contaminated femoral segmental defect model with 6- or 24-hour post-contamination debridement, Shiels et al. demonstrated that 10% wt/wt rifampin in PMMA significantly reduced colony-forming units of *S. aureus* UAMS-1. This was in contrast to lower rifampin loading masses of 1% (wt/wt), rifampin 1% with vancomycin 1.7%, or rifampin 1% with vancomycin 5%. The 10% rifampin-loaded PMMA beads demonstrated burst-release elution kinetics with continued elution over 13 days. Notably, however, this amount of rifampin resulted in incomplete curing of the PMMA, resulting in beads which took between 1 and 2 hours to cure to a final elastic state. In vivo, only 10% rifampin beads prevented bacterial growth on the implants and the PMMA beads themselves, in contrast to the other formulations tested. Recovery of bacterial growth from some of the PMMA beads suggests that this approach can perpetuate infection by providing a surface for bacterial growth. In contrast to the other treatment groups, animals treated with 10% rifampin-loaded PMMA had no signs of clinical infection and significantly lower 14-day bacterial counts from bone, hardware, and tissue when debrided after 6 hours. Delaying debridement for 24 hours worsened these results overall, but nevertheless bone, orthopedic hardware, and tissue appeared to become sterilized in some animals at the 14-day end point. This again suggests an important role for early intervention in contaminated orthopedic trauma to prevent infections, possibly reflecting the role of bacterial biofilms in creating tolerance to antimicrobials. Importantly, a screen for phenotypic rifampin resistance greater than 4 μg/mL found none among recovered isolates. Other delivery methods, to include direct application of antibiotic powder, are being explored. For example, topical placement of vancomycin powder is being used to prevent surgical site infections in many orthopedic surgeries; initial reports demonstrate that deep surgical site infection rate of posterior lateral fusions decreases from 2.6% to 0.2% [41].

Prevention of surgical site infections is very different from treating a biofilm infection. Using a rat open fracture model, the time from bacterial contamination to treatment with topical vancomycin powder had a profound effect on the infection. When antibiotic powder is placed 6 hours after contaminating wounds with

S. aureus, none of the wounds had bacteria over the threshold for infection; however, all of the wounds had high levels of bacteria when treatment was delayed to 24 hours. This was expected given the poor performance of vancomycin against biofilms. Unlike vancomycin, topical placement of rifampin powder was successful with both early and delayed treatment in this model [48]. This approach overcomes the issues of poor release kinetics and incomplete curing using bone cements. Although effective against biofilm-based infections, there were concerns that placement of rifampin powder would delay fracture healing because it is one of the most cytotoxic antibiotics to osteoblasts [49]. Follow-up studies demonstrated that rifampin powder placed in wounds does not negatively affect normal bone healing in a rat segmental defect model [50].

A novel approach to biofilm defeat involves utilizing endogenous chemical signals of biofilms to disperse the colony. Various substances have been reported to disperse mature biofilms [51–53]. We examined the biofilm dispersal and prevention properties of norspermidine, a polyamine compound, on MDRO strains of bacteria (*A. baumannii*, *Klebsiella pneumoniae*, *Pseudomonas aeruginosa*, *S. aureus*) from human clinical infections [54]. We found this substance to have a variable strain- and species-dependent effect on biofilm reduction, with the most pronounced effect on *A. baumannii* by inhibiting motility and reducing the expression of genes encoding quorum-sensing inhibitors. We also observed evidence of significant toxicity in cell cultures and whole tissue explants on histopathology, which may limit consideration of this compound for clinical development. Sanchez et al. utilized a combination of selected D-amino acids (D-Met, D-Phe, D-Trp) as a dispersal agent for biofilms grown from clinical wound strains of *S. aureus* and *P. aeruginosa* [55]. Application of this substance was found to enhance the antibiofilm activity of some currently available antimicrobials, including clindamycin, rifampin, and vancomycin (against *S. aureus* biofilms) and ciprofloxacin and colistin (against *P. aeruginosa* biofilms).

To address the clinical problem of bacterial contamination of synthetic materials necessary for healing critical bone defects (defined as defects that will not heal spontaneously over the lifetime of the organism), Sanchez et al. impregnated D-amino acids into a synthetic polyurethane bone graft [56]. An equimolar mixture of D-Met, D-Pro, and D-Trp significantly reduced bacterial contamination on the scaffold surface in vitro. When implanted in the contaminated femoral segmental defect model, counts of the strong biofilm-former *S. aureus* UAMS-1 were significantly reduced in bone. This serves as a proof-of-concept which suggests that such novel approaches could have utility in the prevention and treatment of clinical infections.

Leveraging its expertise in drug discovery and development, the Walter Reed Army Institute of Research (WRAIR) has contributed to an understanding of combat-related wound infections and explored novel approaches to therapy. Through a collaboration with the Lawrence Livermore National Laboratory, investigators comprehensively evaluated the wound microbiomes of combat casualties using a culture-independent approach based on a massive gene array capable of detecting more than 3500 organisms known to cause infection in vertebrates, as well as a deep-sequencing approach [57]. The most commonly observed microorganism in the human wound samples was *A. baumannii*, found in 23% of samples. Additionally, presence of the pRAY plasmid of *A. baumannii* was significantly associated with the failure of wounds to heal. *Pseudomonas* spp. were detected more frequently in wounds which failed to heal. Interestingly, an inverse association (favoring wound healing) was found with the presence of organisms associated with the gastrointestinal tract [57]. As part of a broader program investigating the utility of phage (virus-like particles which can infect and kill bacteria) for treatment of antibiotic-resistant wound infections, Regeimbal and colleagues showed favorable efficacy of a phage cocktail against a particularly virulent clinical strain of *A. baumannii* (AB5075) in an insect-based bacterial virulence screening model (*Galleria mellonella*). This was translated into a mouse model of wound infection, showing reductions in animal weight loss, wound bioburden, and wound size resulting from the phage therapy. Importantly, the phage cocktail could only infect 10 of 92 screened clinical isolates of *A. baumannii*, illustrating the narrow spectrum of activity which poses a challenge to broad clinical use [58]. In spite of this issue, WRAIR investigators provided a phage cocktail (under FDA approval) which was used to successfully treat a critically ill human patient with a disseminated MDR *A. baumannii* infection who had failed conventional therapies [59]. Phage therapeutics for orthopedic device-related infections are currently being studied.

WRAIR investigators have also pursued a variety of novel small-molecule approaches to address the biofilm component of wound infections. Sambanthamoorthy and colleagues examined the antibiofilm activity of biologically produced surfactants from *Lactobacillus jensenii* and *L. rhamnosus* against MDR strains of *A. baumannii*, *E. coli*, and *S. aureus*. They found significant efficacy in both preventing new biofilms and dispersing established biofilms and observed minimal toxicity against cultured human lung epithelial cells at biofilm-effective concentrations [60]. The same investigators also performed in silico screening of a library of 15,000 compounds for inhibition of the bacterial enzyme diguanylate cyclase (which generates a signaling molecule), of which 250 were tested. Four compounds were found which inhibited biofilm formation by *P. aeruginosa* and *A. baumannii*, including one (LP 3134) which exhibited no cytotoxicity against cultured human keratinocytes [61]. Nine structural derivatives of this compound were subsequently explored for antibiofilm activity against *A. baumannii*, with seven of them effectively reducing biofilm formation on silicone catheters while exerting minimal toxicity to cultured human mammalian cells [62]. Finally, WRAIR investigators examined the activity of the cathelicidin peptide LL-37, and its metabolic fragments, against clinical isolates of MDR *A. baumannii* and their biofilms. While LL-37 and its KS-30 fragment were the most potent at reducing the biofilm, they also appeared to have limiting cytotoxicity. In contrast, the KR-20 fragment showed less efficient killing but was felt to be the most promising therapeutic candidate on the basis of its reduced cytotoxicity [63]. Perhaps these or other novel approaches provided by US military research laboratories can be successfully advanced to clinical development.

Conclusion

Recent military conflicts have included substantial challenges with wound infections, owing to MDRO phenotypes limiting antimicrobial treatment options, and likely involvement from bacterial biofilms contaminating devices implanted to stabilize severe extremity injuries. The US military medical research community responded to these challenges with a variety of in vitro and in vivo studies exploring the biology of these infections as well as potential novel mitigation strategies. The use of animal models of musculoskeletal infection, some of which include implanted devices contaminated with bacteria, has proven to be useful platforms for understanding optimal wound management strategies and screening compounds with promising in vitro antibiofilm activity. As an ideal antibiofilm agent has yet to be identified for clinical development, research in this field should continue in anticipation of wound infections being a significant clinical problem for casualties of future military conflicts.

Disclaimer The opinions or assertions contained herein are the private views of the author and not to be construed as official or as reflecting the views of the Department of the Army or the Department of Defense.

References

1. Jackman, R. P. (2013). Immunomodulation in transfused trauma patients. *Current Opinion in Anaesthesiology, 26*(2), 196–203.
2. Ake, J., et al. (2011). Gram-negative multidrug-resistant organism colonization in a US military healthcare facility in Iraq. *Infection Control and Hospital Epidemiology, 32*(6), 545–552.
3. Landrum, M. L., & Murray, C. K. (2008). Ventilator associated pneumonia in a military deployed setting: The impact of an aggressive infection control program. *The Journal of Trauma, 64*(2 Suppl), S123–S127; discussion S127–8.
4. Scott, P., et al. (2007). An outbreak of multidrug-resistant Acinetobacter baumannii-calcoaceticus complex infection in the US military health care system associated with military operations in Iraq. *Clinical Infectious Diseases, 44*(12), 1577–1584.
5. Akers, K. S., et al. (2014). Biofilms and persistent wound infections in United States military trauma patients: A case-control analysis. *BMC Infectious Diseases, 14*, 190.
6. Owens, B. D., et al. (2007). Characterization of extremity wounds in Operation Iraqi Freedom and Operation Enduring Freedom. *Journal of Orthopaedic Trauma, 21*(4), 254–257.
7. Brown, K. V., Murray, C. K., & Clasper, J. C. (2010). Infectious complications of combat-related mangled extremity injuries in the British military. *The Journal of Trauma, 69*(Suppl 1), S109–S115.
8. Murray, C. K., et al. (2011). Prevention of infections associated with combat-related extremity injuries. *The Journal of Trauma, 71*(2 Suppl 2), S235–S257.
9. Masini, B. D., et al. (2011). Rehospitalization after combat injury. *The Journal of Trauma, 71*(1 Suppl), S98–S102.
10. Cross, J. D., et al. (2012). Return to duty after type III open tibia fracture. *Journal of Orthopaedic Trauma, 26*(1), 43–47.
11. Stinner, D. J., et al. (2010). Return to duty rate of amputee soldiers in the current conflicts in Afghanistan and Iraq. *The Journal of Trauma, 68*(6), 1476–1479.

12. Napierala, M. A., et al. (2014). Infection reduces return-to-duty rates for soldiers with Type III open tibia fractures. *Journal of Trauma and Acute Care Surgery, 77*(3 Suppl 2), S194–S197.
13. Huh, J., et al. (2011). Infectious complications and soft tissue injury contribute to late amputation after severe lower extremity trauma. *The Journal of Trauma, 71*(1 Suppl), S47–S51.
14. Tribble, D. R., et al. (2011). Infection-associated clinical outcomes in hospitalized medical evacuees after traumatic injury: Trauma Infectious Disease Outcome Study. *The Journal of Trauma, 71*(1 Suppl), S33–S42.
15. Weintrob, A. C., et al. (2018). Early infections complicating the care of combat casualties from Iraq and Afghanistan. *Surgical Infections, 19*(3), 286–297.
16. Johnson, E. N., et al. (2007). Infectious complications of open type III tibial fractures among combat casualties. *Clinical Infectious Diseases, 45*(4), 409–415.
17. Yun, H. C., Branstetter, J. G., & Murray, C. K. (2008). Osteomyelitis in military personnel wounded in Iraq and Afghanistan. *The Journal of Trauma, 64*(2 Suppl), S163–S168; discussion S168.
18. Eastridge, B. J., et al. (2009). Impact of joint theater trauma system initiatives on battlefield injury outcomes. *American Journal of Surgery, 198*(6), 852–857.
19. Tribble, D. R., et al. (2018). Osteomyelitis risk factors related to combat trauma open tibia fractures: A case-control analysis. *Journal of Orthopaedic Trauma, 32*, e344.
20. McDonald, J. R., et al. (2018). Infectious complications after deployment trauma: Following wounded United States military personnel into veterans affairs care. *Clinical Infectious Diseases, 67*, 1205.
21. Keeling, J. J., et al. (2008). Short-term outcomes of severe open wartime tibial fractures treated with ring external fixation. *The Journal of Bone and Joint Surgery. American Volume, 90*(12), 2643–2651.
22. Costerton, J. W., Montanaro, L., & Arciola, C. R. (2005). Biofilm in implant infections: Its production and regulation. *The International Journal of Artificial Organs, 28*(11), 1062–1068.
23. Costerton, J. W. (2005). Biofilm theory can guide the treatment of device-related orthopaedic infections. *Clinical Orthopaedics and Related Research*, (437), 7–11.
24. Forsberg, J. A., et al. (2008). Correlation of procalcitonin and cytokine expression with dehiscence of wartime extremity wounds. *The Journal of Bone and Joint Surgery. American Volume, 90*(3), 580–588.
25. Hawksworth, J. S., et al. (2009). Inflammatory biomarkers in combat wound healing. *Annals of Surgery, 250*(6), 1002–1007.
26. Utz, E. R., et al. (2010). Metalloproteinase expression is associated with traumatic wound failure. *The Journal of Surgical Research, 159*(2), 633–639.
27. Brown, T. S., et al. (2011). Inflammatory response is associated with critical colonization in combat wounds. *Surgical Infections, 12*(5), 351–357.
28. Evans, K. N., et al. (2012). Inflammatory cytokine and chemokine expression is associated with heterotopic ossification in high-energy penetrating war injuries. *Journal of Orthopaedic Trauma, 26*(11), e204–e213.
29. Sanchez, C. J., Jr., et al. (2013). Biofilm formation by clinical isolates and the implications in chronic infections. *BMC Infectious Diseases, 13*, 47.
30. Heitkamp, R. A., et al. (2018). Association of enterococcus spp. with severe combat extremity injury, intensive care, and polymicrobial wound infection. *Surgical Infections, 19*(1), 95–103.
31. Cardile, A. P., et al. (2014). Human plasma enhances the expression of Staphylococcal microbial surface components recognizing adhesive matrix molecules promoting biofilm formation and increases antimicrobial tolerance In Vitro. *BMC Research Notes, 7*, 457.
32. Akers, K. S., et al. (2015). Biofilm formation by clinical isolates and its relevance to clinical infections. *Advances in Experimental Medicine and Biology, 830*, 1–28.
33. Penn-Barwell, J. G., Murray, C. K., & Wenke, J. C. (2012). Comparison of the antimicrobial effect of chlorhexidine and saline for irrigating a contaminated open fracture model. *Journal of Orthopaedic Trauma, 26*(12), 728–732.

34. Svoboda, S. J., et al. (2006). Comparison of bulb syringe and pulsed lavage irrigation with use of a bioluminescent musculoskeletal wound model. *The Journal of Bone and Joint Surgery. American Volume, 88*(10), 2167–2174.
35. Owens, B. D., White, D. W., & Wenke, J. C. (2009). Comparison of irrigation solutions and devices in a contaminated musculoskeletal wound survival model. *The Journal of Bone and Joint Surgery. American Volume, 91*(1), 92–98.
36. Barsoumian, A., et al. (2013). In vitro toxicity and activity of Dakin's solution, mafenide acetate, and amphotericin B on filamentous fungi and human cells. *Journal of Orthopaedic Trauma, 27*(8), 428–436.
37. Homeyer, D. C., et al. (2015). In vitro activity of Melaleuca alternifolia (tea tree) oil on filamentous fungi and toxicity to human cells. *Medical Mycology, 53*(3), 285–294.
38. Yabes, J. M., et al. (2017). In Vitro activity of Manuka Honey and polyhexamethylene biguanide on filamentous fungi and toxicity to human cell lines. *Medical Mycology, 55*(3), 334–343.
39. Molinari, R. W., Khera, O. A., & Molinari, W. J., 3rd. (2012). Prophylactic intraoperative powdered vancomycin and postoperative deep spinal wound infection: 1,512 consecutive surgical cases over a 6-year period. *European Spine Journal, 21 Suppl 4*, S476–S482.
40. Pahys, J. M., et al. (2013). Methods to decrease postoperative infections following posterior cervical spine surgery. *The Journal of Bone and Joint Surgery. American Volume, 95*(6), 549–554.
41. Sweet, F. A., Roh, M., & Sliva, C. (2011). Intrawound application of vancomycin for prophylaxis in instrumented thoracolumbar fusions: Efficacy, drug levels, and patient outcomes. *Spine (Phila Pa 1976), 36*(24), 2084–2088.
42. Tennent, D. J., et al. (2016). Time-dependent effectiveness of locally applied vancomycin powder in a contaminated traumatic orthopaedic wound model. *Journal of Orthopaedic Trauma, 30*(10), 531–537.
43. Owens, B. D., & Wenke, J. C. (2007). Early wound irrigation improves the ability to remove bacteria. *The Journal of Bone and Joint Surgery. American Volume, 89*(8), 1723–1726.
44. Lack, W. D., et al. (2015). Type III open tibia fractures: Immediate antibiotic prophylaxis minimizes infection. *Journal of Orthopaedic Trauma, 29*(1), 1–6.
45. Zheng, Z., & Stewart, P. S. (2002). Penetration of rifampin through Staphylococcus epidermidis biofilms. *Antimicrobial Agents and Chemotherapy, 46*(3), 900–903.
46. Liu, C., et al. (2011). Clinical practice guidelines by the Infectious Diseases Society of America for the treatment of methicillin-resistant Staphylococcus aureus infections in adults and children: Executive summary. *Clinical Infectious Diseases, 52*(3), 285–292.
47. Shiels, S. M., et al. (2017). Determining potential of PMMA as a depot for rifampin to treat recalcitrant orthopaedic infections. *Injury, 48*(10), 2095–2100.
48. Shiels, S. M., Tennent, D. J., & Wenke, J. C. (2018). Topical rifampin powder for orthopaedic trauma part I: Rifampin powder reduces recalcitrant infection in a delayed treatment musculoskeletal trauma model. *Journal of Orthopaedic Research, 36*, 3136.
49. Rathbone, C. R., et al. (2011). Effect of various concentrations of antibiotics on osteogenic cell viability and activity. *Journal of Orthopaedic Research, 29*(7), 1070–1074.
50. Shiels, S. M., et al. (2018). Topical rifampin powder for orthopaedic trauma part II: Therapeutic levels of topical rifampin allows for spontaneous bone healing in both sterile and contaminated wounds. *Journal of Orthopaedic Research, 36*, 3142–3150.
51. Bhattacharjee, A., Nusca, T. D., & Hochbaum, A. I. (2016). Rhamnolipids mediate an interspecies biofilm dispersal signaling pathway. *ACS Chemical Biology, 11*(11), 3068–3076.
52. Kumar Shukla, S., & Rao, T. S. (2013). Dispersal of Bap-mediated Staphylococcus aureus biofilm by proteinase K. *Journal of Antibiotics (Tokyo), 66*(2), 55–60.
53. Roizman, D., et al. (2017). In vitro evaluation of biofilm dispersal as a therapeutic strategy to restore antimicrobial efficacy. *Antimicrobial Agents and Chemotherapy, 61*(10).
54. Cardile, A. P., et al. (2017). Activity of norspermidine on bacterial biofilms of multidrug-resistant clinical isolates associated with persistent extremity wound infections. *Advances in Experimental Medicine and Biology, 973*, 53–70.

55. Sanchez, C. J., Jr., et al. (2014). D-amino acids enhance the activity of antimicrobials against biofilms of clinical wound isolates of Staphylococcus aureus and Pseudomonas aeruginosa. *Antimicrobial Agents and Chemotherapy, 58*(8), 4353–4361.
56. Sanchez, C. J., Jr., et al. (2013). Effects of local delivery of D-amino acids from biofilm-dispersive scaffolds on infection in contaminated rat segmental defects. *Biomaterials, 34*(30), 7533–7543.
57. Be, N. A., et al. (2014). Microbial profiling of combat wound infection through detection microarray and next-generation sequencing. *Journal of Clinical Microbiology, 52*(7), 2583–2594.
58. Regeimbal, J. M., et al. (2016). Personalized therapeutic cocktail of wild environmental phages rescues mice from Acinetobacter baumannii wound infections. *Antimicrobial Agents and Chemotherapy, 60*(10), 5806–5816.
59. Schooley, R. T., et al. (2017). Development and use of personalized bacteriophage-based therapeutic cocktails to treat a patient with a disseminated resistant Acinetobacter baumannii infection. *Antimicrobial Agents and Chemotherapy, 61*(10).
60. Sambanthamoorthy, K., et al. (2014). Antimicrobial and antibiofilm potential of biosurfactants isolated from lactobacilli against multi-drug-resistant pathogens. *BMC Microbiology, 14*, 197.
61. Sambanthamoorthy, K., et al. (2014). Identification of small molecules inhibiting diguanylate cyclases to control bacterial biofilm development. *Biofouling, 30*(1), 17–28.
62. Sambanthamoorthy, K., et al. (2015). Modulating Acinetobacter baumannii biofilm development with molecules containing 3,4,5-trimethoxy-N,N′,N′-trimethylbenzohydrazide moiety. *Bioorganic & Medicinal Chemistry Letters, 25*(10), 2238–2242.
63. Feng, X., et al. (2013). The human antimicrobial peptide LL-37 and its fragments possess both antimicrobial and antibiofilm activities against multidrug-resistant Acinetobacter baumannii. *Peptides, 49*, 131–137.

Targeting Biofilms in Orthopedic Infection

Karan Goswami and Javad Parvizi

Abstract While the use of orthopedic implants has transformed the treatment of chronic musculoskeletal diseases such as osteoarthritis, the introduction of foreign materials increases the ability of microbes to cause infection more than 100,000-fold (Elek, Annals of the New York Academy of Sciences. 65:85–90, 1956; Parvizi et al. J Am Acad Orthop Surg. 23:S32-43, 2015). Even when implants are successfully placed without infection, their continued presence predisposes patients to infection years after implantation. The annual cost of infected revision total joint arthroplasty to US hospitals, an example of one of the most common device-associated infections, is projected to exceed $1.62 billion by the year 2020 (Kurtz et al. J Arthroplasty. 27:61–65.e1, 2012). Treatment of prosthesis-associated infections is complex as implants serve as a surface for microbial growth into a resistant biofilm layer. This biofilm layer makes bacteria more difficult to eradicate, facilitates host immune evasion, propagates antimicrobial resistance, and reduces the efficacy of standard antibiotic therapy. Over recent years, numerous strategies have been investigated to prevent, target, and disrupt biofilm on orthopedic implants. We describe the main modes of biofilm-disrupting technology pertinent to orthopedics that have been examined over the last decade – including biofilm localization techniques, implant material modification, bioactive antibacterial coatings, vaccines, bacteriophages, electrical stimulation, and inhibition of quorum sensing. Of note, the success of these novel antibiofilm approaches is currently largely limited to the preclinical setting or early clinical stages. Collaborative efforts between industry, academia, and regulatory authorities are required to fuel translation of this innovation into the clinical arena and ultimately lead to improved patient outcomes.

Keywords Orthopedic · Implants · Biofilm · Translational research · Therapies · Development · Testing

---

K. Goswami (✉) · J. Parvizi
Rothman Institute of Orthopedics at Thomas Jefferson University, Philadelphia, PA, USA

D. L. Williams (ed.), *Targeting Biofilms in Translational Research, Device Development, and Industrial Sectors*,
https://doi.org/10.1007/978-3-030-30667-0_5

Introduction

The Problem of Biofilms in Orthopedics

The surfaces of orthopedic implants are all susceptible to colonization by biofilm-forming microorganisms, whose presence has been reported to play a key role in the pathogenesis of implant-associated infections, such as periprosthetic joint infection (PJI) [4–7]. In terms of pathogenesis, PJI can be initiated through hematogenous spread or by direct seeding via an overlying infection, penetrating trauma, or contamination during surgical implantation of the prosthesis. Regardless of the seeding source or microbial species, the stepwise progression of the infection is dependent upon biofilm formation and maturation [8]. Numerous therapeutic challenges arise as a result of formation of biofilm, particularly as it leads to protection against host immune defense and standard antimicrobial regimens [9].

Biofilm-associated prosthetic infection represents a tremendous burden for patients globally, with massive healthcare cost implications. PJI alone affects between 1% and 2% of all primary total joint arthroplasties and often results in prolonged intravenous antibiotics, additional surgical procedures, longer inpatient stays, significant patient morbidity, and even mortality [10]. Current treatment strategies involve a one- or two-stage approach to revision surgery, with removal of infected components and insertion of local antibiotic-impregnated cement, as well as a course of systematic antimicrobial therapy. However, despite best efforts and considerable ongoing research, treatment success rates are varied and suboptimal [11].

To overcome these problems, novel treatment strategies focusing on disrupting biofilms are being developed [12]. Such antibiofilm strategies hold promise for prevention and improved outcomes of implant-associated orthopedic infections beyond the limitations of current invasive management strategies. This chapter will discuss promising technologies with translational potential for future orthopedic application – including biofilm mapping methods, material modifications, bioactive surface coatings, vaccines, bacteriophages, and inhibition of quorum sensing.

Translational Research

Biofilm Mapping: Detection and Localization

The ability to identify biofilms intraoperatively would be useful in the management of prosthesis-associated infection, particularly in the setting of acute PJI [13]. While it seems intuitive that knowledge of biofilm location would aid in guiding surgical therapy, existing research regarding biofilm mapping is limited and cannot definitely define the clinical importance of this practice for orthopedic infection.

Parry et al. investigated the utility of methylene blue, a disclosing agent traditionally used in dentistry to identify plaque biofilms, for intraoperative detection of biofilms on orthopedic implants. Methylene blue was found to stain *S. epidermidis* biofilm on polyethylene liners, polymethylmethacrylate (PMMA), and Teflon discs in vitro without compromising bacterial viability [14]. However, these experiments were limited to the use of only a single organism's biofilm, *S. epidermidis*, at a much higher density of bacteria than would be expected clinically. Future work is therefore needed to assess the ability of methylene blue to stain physiological levels of biofilm and non-*S. epidermidis* biofilm before it can be translated for clinical application.

Stoodley et al. showed that colored fluorescent proteins can be expressed to directly observe *Pseudomonas aeruginosa* biofilms on 316 L stainless steel screws [15]. Patches of biofilm development were noted on screw shafts and between threads of multiple screws, with no significant pattern of development seen. Confocal laser scanning microscopy has also been shown to aid in biofilm visualization on implant materials and surrounding tissue [16]. However, no focused or systematic analysis exists regarding mapping or formation of the biofilm on specific components or anatomic sites.

The utility of ultrasonication for the detection of biofilms in PJI cases has been explored by Kobayashi et al. and Nguyen et al., who demonstrated that brief exposure of 1–5 minutes of infected components to ultrasonication is effective in detecting bacterial adherence [17, 18]. However, few components were shown to harbor bacteria, and the investigators did not specifically examine for anatomic or component-specific variation.

Material Composition of Orthopedic Components

Another translational avenue is the material composition of orthopedic components, given the possible varying ability of different materials to harbor biofilm formation. The type of alloy used in implants has thus been the focus of many research studies. When testing the ability of some of the more common bacteria in implant infections such as *Staphylococcus aureus*, it was found that bacteria had decreased ability to adhere and create a biofilm layer on titanium versus stainless steel or PMMA [19]. This has been attributed to the ability of titanium to keep bacteria dispersed on the implant surface and therefore leaving bacteria more susceptible to antibiotics [20]. Sheehan et al. compared stainless steel and titanium components using isolated strains of *S. aureus* and *S. epidermidis* in a femoral intramedullary implantation model in rabbits [21]. They noted significantly higher levels of biofilm adherence to stainless steel components within the first 48 hours. Both species showed this preferential growth, with higher levels of adherence reaching nearly 150% on stainless steel compared to titanium. This antimicrobial trend is one reason why titanium alloy has become one of the more popular alloys used in orthopedic implants. Innovations in titanium, including vanadium-free titanium alloys, have been recently investigated by Walkowiak-Przybyło et al. They noted that this specialized alloy exhibited decreased bacterial adherence and biofilm formation than titanium that contains vanadium [22]. Additional avenues of antibiofilm implant modification are discussed and subclassified below.

Intrinsically Bioactive Materials

Bioactive materials that are non-antibiotic compounds with innate antibacterial properties in their structure, such as silver and copper metals, have also been investigated. The antibacterial properties of the metals come from their corrosive properties that result in ion release that can disrupt essential processes of the bacteria such as those in the respiratory chain [23]. These metals can be integrated onto the surface of an orthopedic implant to prevent adhesion of the bacteria.

Silver has been known throughout history for its powerful antimicrobial effects [24, 25]. The mechanism of action is thought to be the formation of reactive oxygen species and active ions that damage bacterial walls and bind to nucleic acids to interrupt bacterial replication [26]. Harrasser et al. [27] studied the antimicrobial effects of silver and have observed significant antimicrobial activity that was positively correlated with its concentration. Silver has shown antibacterial efficacy and biocompatibility when used in combination with coatings of calcium phosphate-hydroxyapatite and ceramics [28]. However, there is concern that the silver layer may influence the metabolic status of adherent cells as well as the properties of the implant in vivo. There is further concern that when the silver release is complete, the implant surface will no longer function as a microbicidal agent. Although rare, there is also the problem of microbial silver resistance and host hypersensitivity to silver ions [29].

Bioactive Antibacterial Coatings and Surface Modification

Development in bioactive antibacterial coatings placed on top of an implant surface has also been a focus of research in recent years. For example, cross-linking the implant surface with a bioactive peptide such as human beta-defensin-3 has been found to reduce the number of bacterial colonies that accumulate on titanium surfaces and is effective against resistant organisms such as methicillin-resistant *Staphylococcus epidermidis* (MRSE) and methicillin-resistant *S. aureus* (MRSA) [30, 31].

Another recently developed antibacterial coating releases nitric oxide that can combine with superoxide to produce peroxynitrite, which has cytotoxic activity against microorganisms [23]. Polymer coating-containing diazeniumdiolates are an example of such a nitric oxide-releasing coating [23, 32]. There are also photoactivated biomaterials that can be activated at certain UV wavelengths to exhibit bactericidal effects such as the anatase TiO_2, which is triggered at UV wavelengths around 385 nm [23].

Hickok et al. examined the role of antibiotic-bonded prostheses in preventing bacterial adhesion to implants, thus reducing the biofilm formation and preventing its ability to harbor bacteria [33]. Vancomycin has typically been used due to its action against gram-positive bacteria, through inhibition of structural bacteria cell wall protein synthesis [34–37]. Multiple other antibiotics have also been explored, such as gentamicin, doxycycline, ceftriaxone, levofloxacin, tetracycline, and other novel agents [33, 38–40]. One method of emitting the antibiotics is through a "controlled-release system" that enables release over a period of multiple days to weeks [33]. These systems are based on biodegradable or non-biodegradable polymers in the form of a prosthetic coating or a sleeve [41, 42]. The alternative method to a controlled-release system is the use of antibiotics that are covalently bound to the prosthesis enabling longer-term action [33]. In animal models, vancomycin that was covalently tethered to a modified titanium plate surface showed no evidence of biofilm formation compared to controls when exposed to *S. aureus* [42]. Furthermore, on immunofluorescence staining, it was found that after 3 months of implantation, vancomycin homogenously covered the surface of the prosthesis and remained stable and active with minimal disturbance of the titanium surface [42].

Nanostructured biomaterials containing compounds such as silver or chitosan also have antibiofilm properties [23]. These nanostructures can also be used to modify the surface properties (e.g., solubility and surface charge) of the implant making it more difficult for bacteria to attach [23, 43]. Numerous reports highlight the antibiofilm and antimicrobial properties of silver nanoparticles with limited host cytotoxicity [24, 44–49]. Aurore et al. found that silver nanoparticles enhanced the bactericidal activity in osteoclasts [50], and antibiofilm effects have been demonstrated in vivo [44, 51]. Kalishwaralal et al. demonstrated that silver nanoparticles at a concentration of 100 nM inhibited >95% of biofilm formation from *S. epidermidis* and *P. aeruginosa* [52]. Bone cement impregnated with silver nanoparticles also significantly reduced biofilm formation compared to standard non-impregnated cement [49], with additional reports suggesting a synergistic effect of silver nanoparticles with antibiotics [53–55]. A key advantage of silver nanoparticle-coated surfaces is the ability to exhibit a continuously controlled release of active agents to the periprosthetic region for a substantial period of time, thus working at both the surface layer and also in the immediate peri-implant environment.

Recently, iodine-supported titanium implants have been shown to reduce bacterial attachment and inhibit biofilm formation [56]. Tsuchiya et al. reported on a cohort of 222 patients with postoperative infection who were treated with iodine-supported implants [57]. At mean 18-month follow-up, all cases of infection were effectively treated, and no host cytotoxicity or adverse thyroid function effects were observed. Similarly, Shirai et al. demonstrated a significant reduction in pin site infection rates by using iodine surface-treated insertion pins and external fixators [58]. In a cohort of 14 revision hip arthroplasties and 16 immunosuppressed primary hip arthroplasties, Kabata et al. showed that iodine-treated hip prostheses remained infection-free at follow-up [59]. No local and systemic toxicity, impaired osteoconductivity, or problems with bony osseointegration were reported in any of these studies.

Vaccines

Vaccination against organisms that commonly infect implants is another area of active research interest. While no vaccine or passive immunization has been approved by the FDA for an orthopedic indication, recently promising studies have explored active and passive strategies for *S. aureus* [60, 61]. The StaphVAX vaccine targeted against *S. aureus* capsular polysaccharides successfully reached phase III clinical trials but was withdrawn when the vaccine's effectiveness in producing immunoglobins against the bacteria decreased to below 30% at 1 year [62]. A separate quadrivalent vaccine against *S. aureus* antigens (targeting glucosaminidase, an ABC transporter lipoprotein, a conserved hypothetical protein, and a conserved lipoprotein) was able to clear 87.5% of biofilm infections in combination with antibiotics versus 22% in those just given the vaccine [63]. Another vaccine against four *S. aureus* antigens has also been shown to be safe and immunogenic in humans in phase I trials [64]. Most recently, the SA4Ag four-antigen vaccine has demonstrated efficacy and safety at beyond 1 year post-immunization in healthy volunteers [65]. This vaccine is currently being further tested in a phase II clinical trial involving spinal fusion patients. There is also the potential for the future development of a vaccine against *Pseudomonas* [66, 67]. However, there are no high-level studies supporting clinical use of the aforementioned vaccines, and further research is needed.

Bacteriophages

The use of lytic bacteriophages, which are natural viruses that infect and destroy bacteria, has recently been explored for eradication of biofilms from orthopedic implants [6, 68]. Lytic phages inject their genetic material into the host bacterial cell, causing bacterial cell lysis, which liberates subsequent new phage particles, and these new particles cause successive infection of additional bacteria in a self-amplifying, exponential pattern. Furthermore, phage therapy appears free of local tissue toxicity or adverse effects, since bacteriophages do not affect eukaryotic cells [69].

Yilmaz et al. found that bacteriophages enhanced the effects of antibiotics in eliminating orthopedic implant infections of MRSA and *P. aeruginosa* in rat models [70]. Ferry et al. injected a local bacteriophage mix during a debridement, antibiotics and implant retention (DAIR) procedure for treatment of an 80-year old patient with relapsing *S. aureus* chronic PJI. This salvage treatment was found to be safe and clinically successful [71].

Although some preclinical and clinical data have demonstrated a good safety profile, as well as promising therapeutic efficacy using bacteriophages for treating orthopedic infections, further clinical research using bacteriophage therapy in patients is required. Current obstacles to bacteriophage translation include the fact that phages are neutralized in the serum and relevant pathogens contain CRISPR-C as immunity against bacteriophage [72]. Phages are also usually bacterial strain-specific; thus, a cocktail of different types of bacteriophages may be necessary to effectively treat a biofilm-mediated infection with this approach.

Bioactive Enzymes

Targeting enzymes that lyse key elements of the biofilm aggregating on orthopedic implants can result in the destruction of the physical integrity of the biofilm matrix. For instance, recombinant human deoxyribonuclease I (rhDNase I) degrades the extracellular DNA component of bacterial biofilms, which is important for cohesion, antimicrobial resistance, and genetic exchange [73]. Similarly, coating of surfaces with dispersin B (DspB) has been found to inhibit >98% of biofilm formation by two clinical strains of *Staphylococcus epidermidis* [74] via exploitation of this natural enzyme's activity against exopolysaccharide biofilm components [23]. In vivo studies have found DspB to have antibiofilm and antibacterial activity against *S. aureus* and *E. coli* when combined with antimicrobial triclosan [75].

Multiple chemotherapeutic agents have also been found to be successful in removing biofilms from implant surfaces. In a recent systematic review, the most successful cytotoxic agent at removing biofilm from contaminated titanium surfaces was citric acid [76]. While there is limited literature regarding these agents, they show promise, and further work is needed to determine their efficacy.

Shockwave Treatment, Electromagnetic Fields, and Electrical Stimulation

Laser and ultrasound-generated shockwave treatment can use mechanical energy to breakup biofilms via disruption of bacterial adhesion. The disrupted biofilm then enables greater exposure of microbes to antibiotic treatments. Kizhner et al. found that around 98% of *P. aeruginosa* biofilms on metallic and plastic medical device surfaces could be removed with between 4 and 10 seconds of laser application. Laser-generated shockwaves were able to break up the biofilm layer into planktonic bacteria amenable to conventional treatment with antibiotics [77]. Similarly, in vivo studies examining 24 hours of continuous ultrasound treatment combined with administration of gentamicin acting on established *E. coli* biofilms demonstrated a significant reduction in viable bacteria [78].

Pulsed electromagnetic field (PEMF) application to bacterial biofilms has also demonstrated antibiofilm effects in vitro and has been found to augment antibiotic treatment efficacy. Pickering et al. applied PEMF to stainless steel pegs infected with biofilms of *S. epidermis* in combination with gentamicin. They reported a 50% reduction in the minimum biofilm inhibitory concentration needed for gentamicin and significant efficacy augmentation after PEMF application [79].

Electrical stimulation of orthopedic implant surfaces also holds promise as a method of biofilm disruption. Ercan et al. anodized and charged nanotubular titanium using15–30 volts of electrical stimulation. They found that *S. aureus* biofilm formation significantly decreased secondary to the formation of fluorine on the surfaces of the anodized titanium [80]. Likewise, electrical polarization of bioceramic hydroxyapatite resulted in a marked reduction in adhesion and proliferation of *S. aureus* and *E. coli* on the positively charged surface [81].

Quorum Sensing Inhibitors

Quorum sensing inhibitors impede the ability of bacteria to communicate with each other, thereby interrupting biofilm development. While there is extensive in vitro and in silica research being conducted to explore this mechanism and anti-quorum sensing molecules, otherwise known as quorum quenching, there are limited in vivo data, and no anti-quorum sensing strategy is currently ready for widespread clinical application. Seven in vivo investigations have been reported during the last 5 years [82–88] with variable experimental strategies. Recent seminal work by Piewngam et al. provides evidence for the elimination of *S. aureus* by probiotic *Bacillus* via inhibition of quorum sensing [89]. They present a detailed molecular mechanism mediated by a class of *Bacillus* lipoproteins, the fengycins, which form the basis for promising future probiotic-based methods of *S. aureus* decolonization and elimination of *S. aureus* infection.

Conclusion

Biofilm-associated periprosthetic infection remains a significant cause of morbidity and mortality for orthopedic patients, with current treatments being limited by their invasive nature and inability to consistently eradicate bacterial biofilm. Several promising treatment modalities have been discussed in this chapter that require further clinical research and trials to bring them into orthopedic practice. Ultimately, a concerted effort from scientists, clinicians, and regulatory authorities is required to work safely and swiftly translate these innovations in order to improve outcomes in patients afflicted with biofilm-associated prosthetic infections.

References

1. Elek, S. D. (1956). Experimental staphylococcal infections in the skin of man. *Annals of the New York Academy of Sciences, 65*, 85–90.
2. Parvizi, J., Alijanipour, P., Barberi, E. F., Hickok, N. J., Phillips, K. S., Shapiro, I. M., et al. (2015). Novel developments in the prevention, diagnosis, and treatment of periprosthetic joint infections. *The Journal of the American Academy of Orthopaedic Surgeons, 23*(Suppl), S32–S43.
3. Kurtz, S. M., Lau, E., Watson, H., Schmier, J. K., & Parvizi, J. (2012). Economic burden of periprosthetic joint infection in the United States. *The Journal of Arthroplasty, 27*, 61–65.e1.
4. Urish, K. L., DeMuth, P. W., Kwan, B. W., Craft, D. W., Ma, D., Haider, H., et al. (2016). Antibiotic-tolerant Staphylococcus aureus biofilm persists on arthroplasty materials. *Clinical Orthopaedics and Related Research, 474*, 1649–1656.
5. Ma, D., Shanks, R. M. Q., Davis, C. M., Craft, D. W., Wood, T. K., Hamlin, B. R., et al. (2017). Viable bacteria persist on antibiotic spacers following two-stage revision for periprosthetic joint infection. *Journal of Orthopaedic Research, 36*, 452–458.
6. Tzeng, A., Tzeng, T. H., Vasdev, S., Korth, K., Healey, T., Parvizi, J., et al. (2015). Treating periprosthetic joint infections as biofilms: Key diagnosis and management strategies. *Diagnostic Microbiology and Infectious Disease, 81*, 192–200.
7. Gbejuade, H. O., Lovering, A. M., & Webb, J. C. (2015). The role of microbial biofilms in prosthetic joint infections. *Acta Orthopaedica, 86*, 147–158.
8. Tande, A. J., & Patel, R. (2014). Prosthetic joint infection. *Clinical Microbiology Reviews, 27*, 302–345.
9. Costerton, J. W., Lewandowski, Z., Caldwell, D. E., Korber, D. R., & Lappin-Scott, H. M. (1995). Microbial biofilms. *Annual Review of Microbiology, 49*, 711–745.
10. Zimmerli, W., Trampuz, A., & Ochsner, P. E. (2004). Prosthetic-joint infections. *The New England Journal of Medicine, 351*, 1645–1654.
11. Tan, T. L., Goswami, K., Fillingham, Y. A., Shohat, N., Rondon, A. J., & Parvizi, J. (2018). Defining treatment success after 2-stage exchange arthroplasty for Periprosthetic joint infection. *The Journal of Arthroplasty, 33*, 3541–3546.
12. George, D. A., Gant, V., & Haddad, F. S. (2015). The management of periprosthetic infections in the future. *The Bone & Joint Journal, 97-B*, 1162–1169.
13. Duque, A. F., Post, Z. D., Lutz, R. W., Orozco, F. R., Pulido, S. H., & Ong, A. C. (2017). Is there still a role for irrigation and debridement with liner exchange in acute periprosthetic total knee infection? *The Journal of Arthroplasty, 32*, 1280–1284.
14. Parry, J. A., Karau, M. J., Kakar, S., Hanssen, A. D., Patel, R., & Abdel, M. P. (2017). Disclosing agents for the intraoperative identification of biofilms on orthopedic implants. *The Journal of Arthroplasty, 32*, 2501–2504.
15. Stoodley, P., Kathju, S., Hu, F. Z., Erdos, G., Levenson, J. E., Mehta, N., et al. (2005). Molecular and imaging techniques for bacterial biofilms in joint arthroplasty infections. *Clinical Orthopaedics and Related Research, 437*, 31–40.
16. Stoodley, P., Nistico, L., Johnson, S., Lasko, L.-A., Baratz, M., Gahlot, V., et al. (2008). Direct demonstration of viable Staphylococcus aureus biofilms in an infected total joint arthroplasty. A case report. *The Journal of Bone and Joint Surgery. American Volume, 90*, 1751–1758.
17. Kobayashi, N., Bauer, T. W., Tuohy, M. J., Fujishiro, T., & Procop, G. W. (2007). Brief ultrasonication improves detection of biofilm-formative bacteria around a metal implant. *Clinical Orthopaedics and Related Research, 457*, 210–213.
18. Nguyen, L. L., Nelson, C. L., Saccente, M., Smeltzer, M. S., Wassell, D. L., & McLaren, S. G. (2002). Detecting bacterial colonization of implanted orthopaedic devices by ultrasonication. *Clinical Orthopaedics and Related Research, 403*, 29–37.
19. Gad, G. F. M., Aziz, A., & Aly Ibrahem, R. (2012). In-vitro adhesion of Staphylococcus spp. to certain orthopedic biomaterials and expression of adhesion genes. *Journal of Applied Pharmaceutical Science, 2*, 145–149.

20. Lauderdale, K. J., Malone, C. L., Boles, B. R., Morcuende, J., & Horswill, A. R. (2010). Biofilm dispersal of community-associated methicillin-resistant Staphylococcus aureus on orthopedic implant material. *Journal of Orthopaedic Research, 28*, 55–61.
21. Sheehan, E., McKenna, J., Mulhall, K. J., Marks, P., & McCormack, D. (2004). Adhesion of Staphylococcus to orthopaedic metals, an in vivo study. *Journal of Orthopaedic Research, 22*, 39–43.
22. Walkowiak-Przybyło, M., Klimek, L., Okrój, W., Jakubowski, W., Chwiłka, M., Czajka, A., et al. (2012). Adhesion, activation, and aggregation of blood platelets and biofilm formation on the surfaces of titanium alloys Ti6Al4V and Ti6Al7Nb. *Journal of Biomedical Materials Research. Part A, 100*, 768–775.
23. Campoccia, D., Montanaro, L., & Arciola, C. R. (2013). A review of the biomaterials technologies for infection-resistant surfaces. *Biomaterials, 34*, 8533–8554.
24. Clement, J. L., & Jarrett, P. S. (1994). Antibacterial silver. *Metal Based Drugs, 1*, 467–482.
25. Nair, L. S., & Laurencin, C. T. (2008). Nanofibers and nanoparticles for orthopaedic surgery applications. *The Journal of Bone and Joint Surgery. American Volume, 90*(Suppl 1), 128–131.
26. Brennan, S. A., Ní Fhoghlú, C., Devitt, B. M., O'Mahony, F. J., Brabazon, D., & Walsh, A. (2015). Silver nanoparticles and their orthopaedic applications. *The Bone & Joint Journal, 97-B*, 582–589.
27. Harrasser, N., Jüssen, S., Banke, I. J., Kmeth, R., von Eisenhart-Rothe, R., Stritzker, B., et al. (2015). Antibacterial efficacy of titanium-containing alloy with silver-nanoparticles enriched diamond-like carbon coatings. *AMB Express, 5*, 77.
28. Shimazaki, T., Miyamoto, H., Ando, Y., Noda, I., Yonekura, Y., Kawano, S., et al. (2010). In vivo antibacterial and silver-releasing properties of novel thermal sprayed silver-containing hydroxyapatite coating. *Journal of Biomedical Materials Research Part B Applied Biomaterials, 92*, 386–389.
29. Hobman, J. L., & Crossman, L. C. (2015). Bacterial antimicrobial metal ion resistance. *Journal of Medical Microbiology, 64*, 471–497.
30. Huang, Q., Yu, H.-J., Liu, G.-D., Huang, X.-K., Zhang, L.-Y., Zhou, Y.-G., et al. (2012). Comparison of the effects of human β-defensin 3, vancomycin, and clindamycin on Staphylococcus aureus biofilm formation. *Orthopedics, 35*, e53–e60.
31. Zhu, C., Tan, H., Cheng, T., Shen, H., Shao, J., Guo, Y., et al. (2013). Human β-defensin 3 inhibits antibiotic-resistant Staphylococcus biofilm formation. *The Journal of Surgical Research, 183*, 204–213.
32. Nablo, B. J., Prichard, H. L., Butler, R. D., Klitzman, B., & Schoenfisch, M. H. (2005). Inhibition of implant-associated infections via nitric oxide release. *Biomaterials, 26*, 6984–6990.
33. Hickok, N. J., & Shapiro, I. M. (2012). Immobilized antibiotics to prevent orthopaedic implant infections. *Advanced Drug Delivery Reviews, 64*, 1165–1176.
34. Lawson, M. C., Bowman, C. N., & Anseth, K. S. (2007). Vancomycin derivative photopolymerized to titanium kills S. epidermidis. *Clinical Orthopaedics and Related Research, 461*, 96–105.
35. Edin, M. L., Miclau, T., Lester, G. E., Lindsey, R. W., & Dahners, L. E. (1996). Effect of cefazolin and vancomycin on osteoblasts in vitro. *Clinical Orthopaedics and Related Research, 333*, 245–251.
36. Parvizi, J., Wickstrom, E., Zeiger, A. R., Adams, C. S., Shapiro, I. M., Purtill, J. J., et al. (2004). Frank Stinchfield Award. Titanium surface with biologic activity against infection. *Clinical Orthopaedics and Related Research, 429*, 33–38.
37. Lawson, M. C., Hoth, K. C., Deforest, C. A., Bowman, C. N., & Anseth, K. S. (2010). Inhibition of Staphylococcus epidermidis biofilms using polymerizable vancomycin derivatives. *Clinical Orthopaedics and Related Research, 468*, 2081–2091.
38. Hart, E., Azzopardi, K., Taing, H., Graichen, F., Jeffery, J., Mayadunne, R., et al. (2010). Efficacy of antimicrobial polymer coatings in an animal model of bacterial infection associated with foreign body implants. *The Journal of Antimicrobial Chemotherapy, 65*, 974–980.
39. Moskowitz, J. S., Blaisse, M. R., Samuel, R. E., Hsu, H.-P., Harris, M. B., Martin, S. D., et al. (2010). The effectiveness of the controlled release of gentamicin from polyelectrolyte multilay-

ers in the treatment of Staphylococcus aureus infection in a rabbit bone model. *Biomaterials, 31*, 6019–6030.
40. Ooi, N., Miller, K., Randall, C., Rhys-Williams, W., Love, W., & Chopra, I. (2010). XF-70 and XF-73, novel antibacterial agents active against slow-growing and non-dividing cultures of Staphylococcus aureus including biofilms. *The Journal of Antimicrobial Chemotherapy, 65*, 72–78.
41. Sinclair, K. D., Pham, T. X., Farnsworth, R. W., Williams, D. L., Loc-Carrillo, C., Horne, L. A., et al. (2012). Development of a broad spectrum polymer-released antimicrobial coating for the prevention of resistant strain bacterial infections. *Journal of Biomedical Materials Research. Part A, 100*, 2732–2738.
42. Stewart, S., Barr, S., Engiles, J., Hickok, N. J., Shapiro, I. M., Richardson, D. W., et al. (2012). Vancomycin-modified implant surface inhibits biofilm formation and supports bone-healing in an infected osteotomy model in sheep: A proof-of-concept study. *The Journal of Bone and Joint Surgery. American Volume, 94*, 1406–1415.
43. Romanò, C. L., Toscano, M., Romanò, D., & Drago, L. (2013). Antibiofilm agents and implant-related infections in orthopaedics: Where are we? *Journal of Chemotherapy, 25*, 67–80.
44. Secinti, K. D., Özalp, H., Attar, A., & Sargon, M. F. (2011). Nanoparticle silver ion coatings inhibit biofilm formation on titanium implants. *Journal of Clinical Neuroscience, 18*, 391–395.
45. Gallo, J., Havranek, V., & Zapletalova, J. (2010). Risk factors for accelerated polyethylene wear and osteolysis in ABG I total hip arthroplasty. *International Orthopaedics, 34*, 19–26.
46. Romanò, C. L., Gala, L., Logoluso, N., Romanò, D., & Drago, L. (2012). Two-stage revision of septic knee prosthesis with articulating knee spacers yields better infection eradication rate than one-stage or two-stage revision with static spacers. *Knee Surgery, Sports Traumatology, Arthroscopy, 20*, 2445–2453.
47. Ionita, D., Grecu, M., Ungureanu, C., & Demetrescu, I. (2011). Antimicrobial activity of the surface coatings on TiAlZr implant biomaterial. *Journal of Bioscience and Bioengineering, 112*, 630–634.
48. Darouiche, R. O. (1999). Anti-infective efficacy of silver-coated medical prostheses. *Clinical Infectious Diseases, 29*, 1371–1377; quiz 1378.
49. Slane, J., Vivanco, J., Rose, W., Ploeg, H.-L., & Squire, M. (2015). Mechanical, material, and antimicrobial properties of acrylic bone cement impregnated with silver nanoparticles. *Materials Science & Engineering. C, Materials for Biological Applications, 48*, 188–196.
50. Aurore, V., Caldana, F., Blanchard, M., Kharoubi Hess, S., Lannes, N., Mantel, P.-Y., et al. (2018). Silver-nanoparticles increase bactericidal activity and radical oxygen responses against bacterial pathogens in human osteoclasts. *Nanomedicine, 14*, 601–607.
51. Qin, H., Cao, H., Zhao, Y., Zhu, C., Cheng, T., Wang, Q., et al. (2014). In vitro and in vivo anti-biofilm effects of silver nanoparticles immobilized on titanium. *Biomaterials, 35*, 9114–9125.
52. Kalishwaralal, K., BarathManiKanth, S., Pandian, S. R. K., Deepak, V., & Gurunathan, S. (2010). Silver nanoparticles impede the biofilm formation by Pseudomonas aeruginosa and Staphylococcus epidermidis. *Colloids and Surfaces. B, Biointerfaces, 79*, 340–344.
53. Wang, J., Li, J., Qian, S., Guo, G., Wang, Q., Tang, J., et al. (2016). Antibacterial surface design of titanium-based biomaterials for enhanced bacteria-killing and cell-assisting functions against periprosthetic joint infection. *ACS Applied Materials & Interfaces, 8*, 11162–11178.
54. Fayaz, A. M., Balaji, K., Girilal, M., Yadav, R., Kalaichelvan, P. T., & Venketesan, R. (2010). Biogenic synthesis of silver nanoparticles and their synergistic effect with antibiotics: A study against gram-positive and gram-negative bacteria. *Nanomedicine, 6*, 103–109.
55. Naqvi, S. Z. H., Kiran, U., Ali, M. I., Jamal, A., Hameed, A., Ahmed, S., et al. (2013). Combined efficacy of biologically synthesized silver nanoparticles and different antibiotics against multidrug-resistant bacteria. *International Journal of Nanomedicine, 8*, 3187–3195.
56. Inoue, D., Kabata, T., Ohtani, K., Kajino, Y., Shirai, T., & Tsuchiya, H. (2017). Inhibition of biofilm formation on iodine-supported titanium implants. *International Orthopaedics (SICOT), 41*, 1093–1099.

57. Tsuchiya, H., Shirai, T., Nishida, H., Murakami, H., Kabata, T., Yamamoto, N., et al. (2012). Innovative antimicrobial coating of titanium implants with iodine. *Journal of Orthopaedic Science, 17*, 595–604.
58. Shirai, T., Watanabe, K., Matsubara, H., Nomura, I., Fujiwara, H., Arai, Y., et al. (2014). Prevention of pin tract infection with iodine-supported titanium pins. *Journal of Orthopaedic Science, 19*, 598–602.
59. Kabata, T., Maeda, T., Kajino, Y., Hasegawa, K., Inoue, D., Yamamoto, T., et al. (2015). Iodine-supported hip implants: Short term clinical results. *BioMed Research International, 2015*, 368124.
60. Søe, N. H., Jensen, N. V., Jensen, A. L., Koch, J., Poulsen, S. S., Pier, G. B., et al. (2017). Active and passive immunization against Staphylococcus aureus periprosthetic osteomyelitis in rats. *In Vivo, 31*, 45–50.
61. Gustin, M.-P., Ohannessian, R., Giard, M., Caillat-Vallet, E., Savey, A., Vanhems, P., et al. (2017). Use of surveillance data to calculate the sample size and the statistical power of randomized clinical trials testing Staphylococcus aureus vaccine efficacy in orthopedic surgery. *Vaccine, 35*, 6934–6937.
62. Shinefield, H., Black, S., Fattom, A., Horwith, G., Rasgon, S., Ordonez, J., et al. (2002). Use of a Staphylococcus aureus conjugate vaccine in patients receiving hemodialysis. *The New England Journal of Medicine, 346*, 491–496.
63. Brady, R. A., O'May, G. A., Leid, J. G., Prior, M. L., Costerton, J. W., & Shirtliff, M. E. (2011). Resolution of Staphylococcus aureus biofilm infection using vaccination and antibiotic treatment. *Infection and Immunity, 79*, 1797–1803.
64. Levy, J., Licini, L., Haelterman, E., Moris, P., Lestrate, P., Damaso, S., et al. (2015). Safety and immunogenicity of an investigational 4-component Staphylococcus aureus vaccine with or without AS03B adjuvant: Results of a randomized phase I trial. *Human Vaccines & Immunotherapeutics, 11*, 620–631.
65. Frenck, R. W., Creech, C. B., Sheldon, E. A., Seiden, D. J., Kankam, M. K., Baber, J., et al. (2017). Safety, tolerability, and immunogenicity of a 4-antigen Staphylococcus aureus vaccine (SA4Ag): Results from a first-in-human randomised, placebo-controlled phase 1/2 study. *Vaccine, 35*, 375–384.
66. de Bruyn, G., Saleh, J., Workman, D., Pollak, R., Elinoff, V., Fraser, N. J., et al. (2016). Defining the optimal formulation and schedule of a candidate toxoid vaccine against Clostridium difficile infection: A randomized Phase 2 clinical trial. *Vaccine, 34*, 2170–2178.
67. Döring, G., Meisner, C., Stern, M., & Flagella Vaccine Trial Study Group. (2007). A double-blind randomized placebo-controlled phase III study of a *Pseudomonas aeruginosa* flagella vaccine in cystic fibrosis patients. *Proceedings of the National Academy of Sciences of the United States of America, 104*, 11020–11025.
68. Akanda, Z. Z., Taha, M., & Abdelbary, H. (2018). Current review-The rise of bacteriophage as a unique therapeutic platform in treating peri-prosthetic joint infections. *Journal of Orthopaedic Research, 36*, 1051–1060.
69. Chhibber, S., Kaur, T., & null, S. K. (2013). Co-therapy using lytic bacteriophage and linezolid: Effective treatment in eliminating methicillin resistant Staphylococcus aureus (MRSA) from diabetic foot infections. *PLoS One, 8*, e56022.
70. Yilmaz, C., Colak, M., Yilmaz, B. C., Ersoz, G., Kutateladze, M., & Gozlugol, M. (2013). Bacteriophage therapy in implant-related infections: An experimental study. *The Journal of Bone and Joint Surgery. American Volume, 95*, 117–125.
71. Ferry, T., Leboucher, G., Fevre, C., Herry, Y., Conrad, A., Josse, J., et al. Salvage debridement, antibiotics and implant retention ("DAIR") with local injection of a selected cocktail of bacteriophages: Is it an option for an elderly patient with relapsing Staphylococcus aureus prosthetic-joint infection? *Open Forum Infectious Diseases, 5*, ofy269. [Internet]. 2018 [cited 2019 Jan 5]; Available from: http://academic.oup.com/ofid/article/5/11/ofy269/5144083.
72. Gasiunas, G., Sinkunas, T., & Siksnys, V. (2014). Molecular mechanisms of CRISPR-mediated microbial immunity. *Cellular and Molecular Life Sciences, 71*, 449–465.

73. Kaplan, J. B., LoVetri, K., Cardona, S. T., Madhyastha, S., Sadovskaya, I., Jabbouri, S., et al. (2012). Recombinant human DNase I decreases biofilm and increases antimicrobial susceptibility in staphylococci. *The Journal of Antibiotics, 65*, 73–77.
74. Pavlukhina, S. V., Kaplan, J. B., Xu, L., Chang, W., Yu, X., Madhyastha, S., et al. (2012). Noneluting enzymatic Antibiofilm coatings. *ACS Applied Materials & Interfaces, 4*, 4708–4716.
75. Darouiche, R. O., Mansouri, M. D., Gawande, P. V., & Madhyastha, S. (2009). Antimicrobial and antibiofilm efficacy of triclosan and DispersinB combination. *The Journal of Antimicrobial Chemotherapy, 64*, 88–93.
76. Ntrouka, V. I., Slot, D. E., Louropoulou, A., & Van der Weijden, F. (2011). The effect of chemotherapeutic agents on contaminated titanium surfaces: A systematic review. *Clinical Oral Implants Research, 22*, 681–690.
77. Kizhner, V., Krespi, Y. P., Hall-Stoodley, L., & Stoodley, P. (2011). Laser-generated shockwave for clearing medical device biofilms. *Photomedicine and Laser Surgery, 29*, 277–282.
78. Rediske, A. M., Roeder, B. L., Brown, M. K., Nelson, J. L., Robison, R. L., Draper, D. O., et al. (1999). Ultrasonic enhancement of antibiotic action on Escherichia coli biofilms: An in vivo model. *Antimicrobial Agents and Chemotherapy, 43*, 1211–1214.
79. Pickering, S. A., Bayston, R., & Scammell, B. E. (2003). Electromagnetic augmentation of antibiotic efficacy in infection of orthopaedic implants. *Journal of Bone and Joint Surgery. British Volume (London), 85*, 588–593.
80. Ercan, B., Kummer, K. M., Tarquinio, K. M., & Webster, T. J. (2011). Decreased Staphylococcus aureus biofilm growth on anodized nanotubular titanium and the effect of electrical stimulation. *Acta Biomaterialia, 7*, 3003–3012.
81. Ueshima, M., Tanaka, S., Nakamura, S., & Yamashita, K. (2002). Manipulation of bacterial adhesion and proliferation by surface charges of electrically polarized hydroxyapatite. *Journal of Biomedical Materials Research, 60*, 578–584.
82. Chatterjee, M., D'Morris, S., Paul, V., Warrier, S., Vasudevan, A. K., Vanuopadath, M., et al. (2017). Mechanistic understanding of Phenyllactic acid mediated inhibition of quorum sensing and biofilm development in Pseudomonas aeruginosa. *Applied Microbiology and Biotechnology, 101*, 8223–8236.
83. Cho, Y.-J., Song, H. Y., Ben Amara, H., Choi, B.-K., Eunju, R., Cho, Y.-A., et al. (2016). In vivo inhibition of porphyromonas gingivalis growth and prevention of periodontitis with quorum-sensing inhibitors. *Journal of Periodontology, 87*, 1075–1082.
84. Durai, S., Vigneshwari, L., & Balamurugan, K. (2013). Caenorhabditis elegans-based in vivo screening of bioactives from marine sponge-associated bacteria against Vibrio alginolyticus. *Journal of Applied Microbiology, 115*, 1329–1342.
85. Husain, F. M., Ahmad, I., Baig, M. H., Khan, M. S., Khan, M. S., Hassan, I., et al. (2016). Broad-spectrum inhibition of AHL-regulated virulence factors and biofilms by sub-inhibitory concentrations of ceftazidime. *RSC Advances, 6*, 27952–27962.
86. Ivanova, K., Fernandes, M. M., Francesko, A., Mendoza, E., Guezguez, J., Burnet, M., et al. (2015). Quorum-quenching and matrix-degrading enzymes in multilayer coatings synergistically prevent bacterial biofilm formation on urinary catheters. *ACS Applied Materials & Interfaces, 7*, 27066–27077.
87. Rowley D., Zhao W., Yuan T., Dao C., Sohn S., Gomez Chiarri M., et al. Mechanisms of microbe-microbe-host interactions in a probiont-pathogen-bivalve model. Planta Medica[Internet], 2015; 81. Available from: http://www.embase.com/search/results?subaction=viewrecord&from=export&id=L71958553.
88. Srinivasan, R., Mohankumar, R., Kannappan, A., Karthick Raja, V., Archunan, G., Karutha Pandian, S., et al. (2017). Exploring the anti-quorum sensing and antibiofilm efficacy of phytol against Serratia marcescens associated acute pyelonephritis infection in Wistar rats. *Frontiers in Cellular and Infection Microbiology, 7*, 498.
89. Piewngam, P., Zheng, Y., Nguyen, T. H., Dickey, S. W., Joo, H.-S., Villaruz, A. E., et al. (2018). Pathogen elimination by probiotic Bacillus via signalling interference. *Nature, 562*, 532.

Translation of Antibiofilm Technologies to Wounds and Other Clinical Care

Matthew Myntti

Abstract Nearly 80% of global bacterial infections are associated with biofilm bacteria (Joo, Otto, Chem Biol 19:1503–1513, 2012). In contrast to planktonic bacteria, biofilms are a complex, organized bacterial community possessing a sophisticated protective armor, in the form of the extracellular polymeric substance (EPS), which acts as a robust defense mechanism against eradication. Chronic biofilm infections affect 17 million people annually, and approximately 550,000 people die as a result of their chronic infections (Wolcott et al J Wound Care 19:45–50, 2010). The challenge with biofilm-related infections is that they cannot be adequately confirmed via diagnostic tests in the clinical setting, and, more importantly, they are intrinsically resistant to host immunity, antibiotics, and biocides. This renders current therapeutic options inadequate to successfully eradicate the infection. Next Science™ has applied novel material science methods to combat biofilm through its innovative Xbio™ technology. Xbio technology, which includes the proprietary product, BlastX™, works by disrupting the biofilm matrix and creating an environment that compromises the biofilm's structural integrity. In doing so, the EPS can be broken down and removed, thereby allowing the pathogens within the environment to be targeted and preventing the biofilm's reformation.

Keywords Wound · Biofilm · Infection · BlastX · Topical · Healing

Since BlastX is considered a combination product, a medical device with a drug component, there were some regulatory challenges in navigating the FDA clearance pathway. BlastX was first submitted to the FDA as an OTC device with limited and standard OTC claims. Once further data was obtained, Next Science submitted a second submission and received clearance for the use of BlastX on more chronic wounds by prescription only.

M. Myntti (✉)
Next Science®, Sydney, NSW, Australia
e-mail: mmyntti@nextscience.org

D. L. Williams (ed.), *Targeting Biofilms in Translational Research, Device Development, and Industrial Sectors*,
https://doi.org/10.1007/978-3-030-30667-0_6

Normal wound physiology goes through four different steps [3]: hemostasis, inflammation, proliferation, and remodeling. Hemostasis takes place seconds to hours after the initial injury. Inflammation can include increased vasodilation and vasopermeability. This can lead to increased exudate, a release of cytokines and growth factors, immune cell recruitment, and, finally, bacterial clearance. The inflammation process typically occurs over a period of hours to days. Proliferation of the wound begins in days to weeks, provided the inflammation is controlled and no infection is present. However, if bacteria infiltrate the wound, a microbial infection can result and interrupt the healing process.

According to the US National Institutes of Health, biofilms account for over 80 percent of microbial infections in the human body [4]. Research has demonstrated that 80–90% of all chronic wounds contain microorganisms protected by biofilms (Fig. 1) [5]. Chronic infections are defined as wounds that take more than 12 weeks to heal, and research states that 70% of wounds worldwide fall under this definition [6]. Chronic biofilm infections can affect every organ system in the human body, including the skin [7]. Approximately 17 million people annually are affected by chronic biofilm infections, and approximately 550,000 people will die each year as a result [2].

The rising prevalence of antibiotic-resistant organisms, particularly within hospitals, is a contributing factor to the prevalence of chronic infections. Antibiotic-resistant organisms and their complications are responsible for more than two million hospital-acquired infections at a cost of $30.5 billion [8]. As discussed, healing for these infections can be routinely delayed by the introduction of microorganisms while the wound remains inflamed. Particularly at risk are those affected by diabetes and vascular disease, where explosive infected numbers have led to a rise in untreatable chronic wounds. This results in an increased burden that negatively impacts the patients' quality of life [9].

Collectively, these chronic wounds significantly contribute to morbidity, mortality, and increased healthcare expenditures [10].

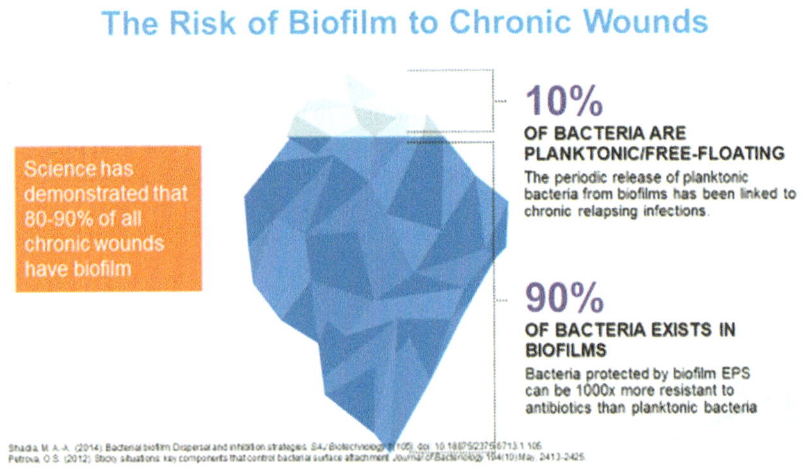

Fig. 1 Biofilm Risk to Chronic Wounds. (Image courtesy of Next Science®)

Bacteria exist in two essential forms: free floating (planktonic) and anchored/sessile (biofilms, spores). While planktonic bacteria are well understood and relatively easy to kill, biofilms pose a unique challenge. Biofilms are surface-adhering bacteria that are encased and defended by a glycocalyx, also known as an extracellular polymeric surface (EPS). This EPS begins to form after the bacteria secrete a sticky gel that protects them from initial eradication. Polymers inside the gel then become cross-linked by metallic bonds to strengthen the structure's integrity and form the backbone of the extracellular polymeric surface. Once metallic bonds become established, the biofilm converts to an insoluble capsular environment that interacts with the host for bacterial growth, mutation, and proliferation. Ninety percent of the bacteria are enveloped within the structure, leaving less than 10% of free-floating bacteria in a wound. The resulting structure is mechanically resistant because metallically bonded polymers anchor the extracellular polymeric structure (EPS), preventing it from being washed off or eradicated by current treatment protocols (Fig. 2).

The EPS acts as a key mechanism in protecting the underlying pathogens by blocking large molecules such as antimicrobials, antibodies, and inflammatory cells from invading. Similarly, its biofilm matrixes act as diffusion barrier to small molecules like antibiotics, safeguarding it from extermination by conventional means [11]. Biofilm matrixes have also developed a mechanism for a subpopulation to become metabolically quiescent, i.e., to hibernate [12, 13]. Furthermore, the EPS exhibits cooperative protective effects. Some species of bacteria can assist others to attach and incorporate into the biofilm (quorum sensing) [14]. The overall effect of these mechanisms is to create a robust, well-defended bacterial community that thrives in spite of elimination efforts.

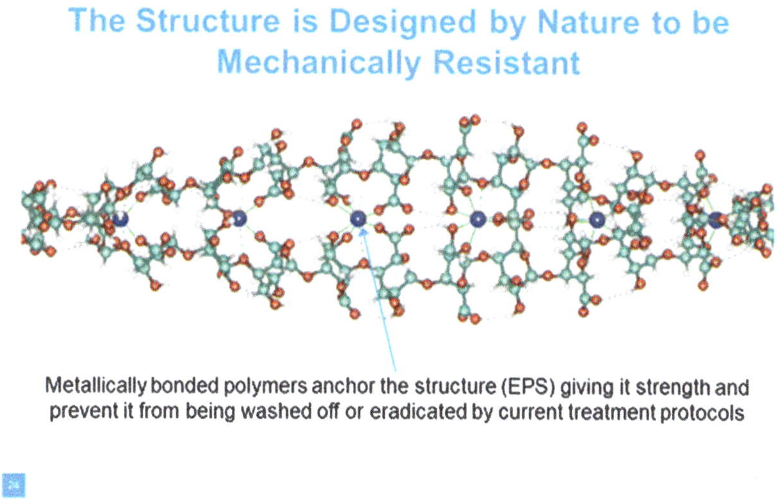

Fig. 2 Extracellular Polymeric Structure of Biofilms. (Image Courtesy of Next Science®)

Antimicrobial drugs are the current mainstay treatment for the management of an acute bacterial infection. Antimicrobials target essential components of bacterial metabolism through the inhibition of cell wall synthesis, cell membrane function, protein synthesis, RNA synthesis, and DNA synthesis. Their primary action mechanism affects bacteria which are metabolically active during a synthesis process of active replication. If at any time bacterial cells become quiescent, or metabolically inactive, they become resistant to most antimicrobials [15]. Therefore, biofilms, with their ability to become metabolically quiescent, are intrinsically resistant to antimicrobials. The biofilms' genetic mechanisms facilitate modification to the antimicrobial target in the form of decreased uptake, efflux pumps, modulation of metabolic pathways, and conferred resistance. Additional functional mechanisms involve modifications to the antimicrobial molecule, prevention of target access, bypass of target sites, or global cell adaption and resistance.

Biofilm bacteria exhibit up to 1000-fold more antimicrobial resistance when compared to planktonic bacteria. Various protective mechanisms render current therapeutic options inadequate to successfully eradicate the infection. Furthermore, the treating clinician often lacks definitive diagnostic data to confirm the presence of biofilm, making the decision to remove infected hardware and tissues, and to treat with antimicrobial agents even more difficult. The decision involves balancing the relative risks of treating or not treating the infection versus exposing a patient to the potential adverse effects of the available treatment strategies.

Current treatment strategies for chronic wound infections generally involve the use of topical antimicrobial dressings as well as local debridement. Debridement breaks biofilm into smaller colonies but does not entirely remove it and may spread the biofilm to other wound regions. Therefore, debridement can amplify the infection, spreading it more aggressively and causing it to undergo reformation faster than on its own. To mitigate these side effects, debridement is generally followed by a topical antimicrobial for highest effectiveness [16]. However, in the context of biofilm-based infections, dosages of antimicrobial drugs up to 500–1000 times the minimum inhibitory concentration are often required. Even if such concentrated dosages were to be administered, they would still be unsuccessful at completely eradicating the infection [22]. An optimal treatment for a biofilm infection should include the use of an antibiofilm agent in addition to the current strategies [17]. A targeted antibiofilm approach is necessary to disrupt and degrade the EPS matrix of the biofilm, target the bacteria for destruction, and prevent biofilm reformation in the wound [18, 19].

Next Science is leading a paradigm shift with a unique, unprecedented approach to eliminating both biofilm bacteria and planktonic bacteria with a proprietary, nontoxic technology that disrupts the biofilm's extracellular polymeric substance (EPS) matrix and makes the bacteria within the biofilm more vulnerable to attack by antimicrobials, antibiotics, and the body's natural immune defenses. This patented Xbio™ technology reduces the bacterial load which, in turn, helps to reduce the overall use of antibiotics (Fig. 3). More importantly, it has shown no known evidence of bacterial resistance [20].

New Therapy Directly Targets the Biofilm Structure

EXPOSE TREAT & PREVENT

Description	Deconstruct EPS	Present Treatment Guidelines			
	Dissolves Biofilm Structure	Broad antimicrobial spectrum	High tissue compatibility	No microbial resistance	Sustained biofilm reformation barrier effects
Dry dressing	NA	NA	✓	NA	X
Targeted antibiotics	No	✓	✓	Variable	X
Topical antimicrobials	No	✓	Variable	Variable	Variable
Outpatient sharp debridement	No	✓	Non-selective	✓	X
Biofilm disruption and microbial lysis	Yes	✓	✓	✓	✓

Wolcott R. *J Wound Care*. 2015;24(8):366-371. Kim P, et al. *Wounds*. 2018;30(5):114-119.
Snyder R, et al. *Wounds*. 2017;Supplement 29(6 suppl);S1-S17.

Fig. 3 Efficacy of BlastX™ compared to traditional standards of care. (Image courtesy of Next Science®)

Next Science's Xbio uses proprietary composition-of-matter patents that contain technology to physically break down the biofilm's protective structures (Fig. 4). The exposure and eradication of the formerly enveloped bacteria are achieved by the technology's induced cell lysis.

Next Science has created BlastX™, an antimicrobial wound gel designed to facilitate natural wound healing. The use of the hydrogel on a wound creates a moist environment that reduces the buildup of necrotic tissue caused by apoptosis and enables the body's natural wound healing process to take place. The moist environment created by the gel promotes granulation, epithelization, and autolytic debridement. The moist environment also prevents tissue dehydration and cell death, increases angiogenesis, and increases the breakdown of dead tissue and fibrin [21].

The gel additionally prevents bacterial growth and the formation of biofilms when applied to fresh wounds by preventing the bacteria from passing through the gel into the wound. BlastX is a topical polyethylene glycol-based hydrogel that disrupts and eliminates biofilms that become enveloped in the gel. This occurs largely by degrading the biofilm's EPS matrix through removal of the metallic bonds in the EPS via chelation and hydrolysis. The hyperosmolar wound gel, and its contained surfactant, enables cell wall lysis, resulting in destruction of the microorganisms that were formerly protected by the biofilm's EPS.

The citric acid in the gel binds to the biofilms' metallic bonds, while the sodium citrate buffers the solution to a pH of 4. This allows the citric acid to attach and remove the metallic bonds that hold the EPS structure together and releases the polymers. Sodium molecules split off and cap the free polymer ends. The remaining sodium citrate molecules are then converted to citric acid. This conversion prevents

Fig. 4 Four ingredients in Xbio™ technology. (Image courtesy of NextScience®)

the polymer from reattaching and replenishes the original citric acid that was depleted in breaking the metallic bonds, thereby sustaining the chelation process through buffering. The pathogens are destroyed when sodium citrate and citric acid in the gel mixture produce an osmotic pressure distending the bacterial cell wall. Aiding in cell lysis, the benzalkonium chloride surfactant then attaches to a protein in the cell wall and removes it. BlastX prevents the recolonization of the biofilms' EPS structure by preventing the bacteria from passing through the gel for biofilm regrowth.

Specifically, once the biofilm enters the gel environment, the Next Science technology dissolves the slime layer permitting direct contact with individual bacteria. Typically, the RNA/proteins in the biofilm's EPS deactivate treatment chemicals before they reach the bacteria. Next Science technology overwhelms these entities, ensuring that critical conditions for lysis are maintained throughout treatment (Fig. 5). Lysis is nondiscriminatory, effective against both gram-positive and gram-negative strains of bacteria, and active, downregulated, and persister cells. The bacteria have no resistance mechanism to cell lysis.

BlastX has been studied extensively to quantify its effectiveness on creating an ideal healing environment for chronic wounds and eliminating robust biofilms. Tests of Suspension Time Killing show that BlastX is effective against a broad range of bacteria and selective fungi, including *C. albicans* and *A. brasiliensis*. A study led by Montana State University demonstrated that BlastX has a nearly six times higher log reduction from control than leading wound gels SilvaSorb and Microcyn, based on a 24 -hour contact time and an 8-log control.

In addition, the applications of BlastX have also been evaluated in vivo. Research conducted at Texas Tech studied the infection reduction in 24-hour biofilm growth with LUX-modified *S. aureus* and *P. aeruginosa* bacteria. Twenty-four hours after the first BlastX application, rats were shown to have significant reduction in infection rates compared to the control. Similarly, WuXi modeled infected rats wound

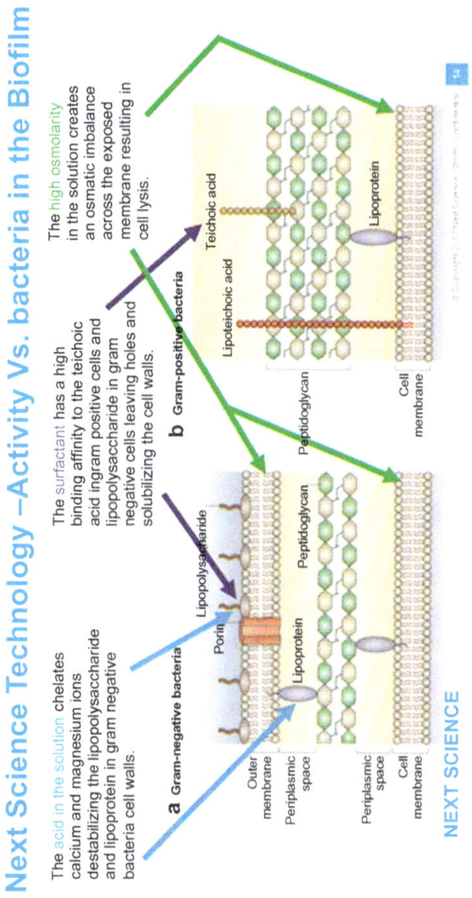

Fig. 5 Xbio™ process of biofilm destruction. (Image courtesy of Next Science®)

size over time (rats infected with *S. aureus*), modeling BlastX's ability to provide a wound healing environment, which resulted in a decrease in wound size at faster rates than the control in the first 7 days of healing. At day 7, wounds covered with BlastX had reduced in size to below 20% of the original area compared to the control's reduction to approximately 55% the original area. This coincided with a reduction in the CFU of bacteria recovered from the wounds, with the bacterial counts reduced from 4.3 log for the control animals to 0.7 log for the BlastX treated rats.

A clinical trial by the Mayo Clinic reviewed the efficacy of BlastX at creating a wound environment which enabled the natural reduction in the size of wounds in human patients. In a 12-week random trial with 43 participants, BlastX was shown to provide a wound environment that resulted in three times the area reduction of chronic wounds over a broad-spectrum antimicrobial ointment. In addition, patients saw a 205% relative increase in wound closure when the wound was covered with BlastX instead of a broad-spectrum antimicrobial ointment. Similarly, a study by Wolcott found that, over a period of 4 weeks, 45 subjects with chronic wounds saw 1.5 times more effective wound closure and 2 times more effective wound area reduction than the standard practice of care.

Next Science currently has four active generations of solution/gel technologies developed: BlastX, Bactisure™ for surgical lavage, Next Science Acne Gel (NAG), and TorrentX [22]. Within these generations, our Intellectual Property covers broad ranges of chemicals and solution properties. This allows Next Science to tailor formulations for specific use conditions, anatomical area, application time frame, and toxicity.

Because the Next Science technology is targeted to attack prokaryotic structures (bacteria and biofilms), they are nontoxic for use on eukaryotic tissues. The pH of Next Science solutions and gels are not hazardous to mammalian tissue. Cells are quite resistant to negative effects of osmolarity due to decreased permeability and the body's ability to normalize the osmolarity from the non-exposed surfaces. There is broad evidence showing that cationic surfactants at low to moderate concentrations are safe for human use. Proteins on the surface of the bacteria are susceptible to binding with cationic surfactants. The solvents used in Next Science products are already used within patients and are used at low concentrations in these products. The enzymes used in Next Science products are commonly present in the human body and pose no toxicity concerns.

Since the Xbio technology is considered to be a combination product, a medical device with a drug component, it required a different path for FDA regulation than current drug-based treatments. For a drug to obtain FDA approval, it must undergo clinical testing and then be submitted to the FDA's Center for Drug Evaluation and Research (CDER). This process can take years and be quite costly. A medical device is approved through the FDA's Center for Devices and Radiological Health (CDRH). Depending on the classification, a device is either cleared or approved for sale. The BlastX device was considered a moderate risk device which required a submission to show substantial equivalence through the FDA's 510(k) process. A 510(k) submission must demonstrate that the device is substantially equivalent to another device legally in commercial distribution in the United States: (1) before May 28, 1976 or (2) to a device that has been determined by FDA to be substantially equivalent [23].

According to the FDA, a combination product is defined as, "a product comprised of two or more regulated components (i.e., drug/device, biologic/device, drug/biologic, or drug/device/biologic) that are physically, chemically, or otherwise combined or mixed and produced as a single entity [often referred to as a "single-entity" combination product]" [24] The Xbio technology is a combination product because BlastX is a wound dressing used to cover and protect the wound, giving it the properties of a device, but it also contains benzalkonium chloride, an antimicrobial agent, which constitutes the drug portion. BlastX was regulated by CDRH due to its primary mode of action being achieved by the device activities of the wound dressing.

BlastX was originally designed with the OTC monographs in mind. OTC monographs allow the marketing of drug products without the requirement for a New Drug Application (NDA), provided specific limitations are placed on the product. The OTC monographs currently allow for a 1:750 (0.13%) use concentration of benzalkonium chloride to be marketed under the category of "skin protectant." The PEG and buffers in the BlastX gel are all accepted as inactive ingredients for US drug products. As such, BlastX could have been marketed as an OTC drug product. The FDA has been moving away from the use of OTC monographs for wound dressings and so Next Science took the next step to submit BlastX to the CDRH division of FDA as a combination wound dressing with an antimicrobial agent. The initial submission was to gain clearance for the same indications that were used with the OTC monographs. Once further data was obtained, Next Science submitted a second submission for the prescription only use of BlastX on more chronic wounds.

For future projects, Next Science will continue to evaluate the appropriate regulatory pathway for each of its new products. Some technologies will most likely be drugs, which will go to the FDA's CDER, while others might be designated as new devices. These new devices would require either a premarket application (PMA) or a de novo application for the establishment of a new device type along with the classification, regulation, and necessary controls and product code. The de novo process is an option for lower-risk devices, and once approved, a de novo device can then serve as a predicate for new medical devices where appropriate to the 501(k) process [25].

Next Science has created a rapid-acting technology, providing options that have superior efficacy against both planktonic and biofilm bacterial forms. Xbio™ is gentle, with low toxicity and a favorable environmental impact. We are at the forefront of addressing the growing problem of biofilm-caused antimicrobial resistance.

Disclaimer Dr. Myntti has financial interest in Next Science and the technologies discussed.

References

1. Joo, H. S., & Otto, M. (2012). Molecular basis of in-vivo biofilm formation by bacterial. *Chemistry & Biology, 19*(12), 1503–1513.
2. Wolcott, R., Rhoads, D., Bennett, M., Wolcott, B., Gogokhia, L., Costerton, J., & Dowd, S. (2010). Chronic wounds and the medical biofilm paradigm. *Journal of Wound Care, 19*(2), p45–p50.
3. Landén, N. X., & Li, D. (2016). Mona Ståhle.Transition from inflammation to proliferation: a critical step during wound healing. *Cellular and Molecular Life Sciences, 73*(20), 3861–3885.
4. Attinger, C., & Wolcott, R. (2012). Clinically addressing biofilm in chronic wounds. *Advances in Wound Care, 1*(3), 127–132.
5. Aziz-Abdel, M. S. (2014). Bacterial biofilm: Dispersal and inhibition strategies. *SAJ Biotechnology, 1*, 105. https://doi.org/10.18875/2375-6713.1.105.
6. Frykberg, R. G., & Banks, J. (2015). Challenges in the Treatment of Chronic Wounds. *Advances in Wound Care, 4*(9), 560–582. https://doi.org/10.1089/wound.2015.0635.
7. Flemming, H. C., & Wingender, J. (2010). The Biofilm Matrix. Nature Reviews Microbiology, 8, 623–633.
8. Performance of the Massachusetts Health Care System Series: A focus on provider quality. CHIA Center for Health Information and Analysis, chiamass.gov.
9. Bjarnsholt, T., Kirketerp-Møller, K., Jensen, P. Ø., et al. (2008). Why chronic wounds will not heal: a novel hypothesis. *Wound Repair and Regeneration, 16*(10), 2–10.
10. Sen, C., Gordillo, G., Roy, S., Kirsner, R., Lambert, L., Hunt, T., Gottrup, F., Gurtner, G., & Longaker, M. (2009). Human skin wounds: A major and snowballing threat to public health and economy. *Wound Repair and Regeneration, 17*(6), 763–771.
11. Hall-Stoodley, L., & Stoodley, P. (2009). Evolving concepts in biofilm infections. *Cell Microbiol, 11*(7), 1034–1043.
12. Brooun, A., Liu, S., & Lewis, K. (2000). A dose response study of antibiotic resistance in Pseudomonas aeruginosa biofilms. *Antimicrobial Agents and Chemotherapy, 44*(3), 640–646.
13. Lewis, K. (2007). Persister cells, dormancy and infectious disease. *Nature Reviews. Microbiology, 5*(1), 48–56.31.
14. Davies, D. G., Parsek, M. R., Pearson, J. P., et al. (1998). The involvement of cell-to-cell signals in the development of a bacterial biofilm. *Science, 280*(5361), 295–298.
15. Carroll, K., Hobden, J., Miller, S., Morse, S., Mietzner, T., Detrick, B., Mitchell, T., McKerrow, J., & Sakanari, J. (2016). *Jawetz, Melnick, & Adelbergs medical microbiology* (27th ed.). Shenzen: McGraw-Hill.
16. Kim et al. (2018). *Wounds*, 30(5), 114–119. Epub 2018 February 23 (Biofilms Made Easy (wounds international) – Phillips; Skin and Wound Care 2013 – Kevin Wu; Wound Education Update Feb 2014 – Wounds UK Best Practice Statement)
17. Snyder, R. J., Bohn, G., Hanft, J., Harkless, L., Kim, P., Lavery, L., Schultz, G., & Wolcott, R. (2017). Wound bioflm: current perspectives and strategies on biofilm disruption and treatments. *Wounds, 29*(6), S1–S17.
18. Snyder, R. J., Bohn, G., Hanft, J., Harkless, L., Kim, P., Lavery, L., Schultz, G., & Wolcott, R. (2017). Wound Biofilm: Current Perspectives and Strategies on Biofilm Disruption and Treatments. *Wounds, 29*(6), S1–S17.
19. Ryder, M. (2005). Catheter related infections – It's all about biofilms. *Topics in Advanced Practice Nursing eJournal, 5*(3), 1–6.
20. Data on file. Center for Biofilm Engineering at Montana State University. Next Science Report TR-10-12-004.
21. Mallefet, P., & Dweck, A. C. (2008). Mechanisms involved in wound healing. *Biomedical Scientist, 52*(7), 609.
22. Next Science website. http://www.nextscience.com
23. Code of Federal Regulations. https://www.fda.gov/medicaldevices/deviceregulationandguidance/overview/default.htm

24. FDA.gov https://www.fda.gov/combinationproducts/aboutcombinationproducts/ucm101496.htm#CP
25. Evaluation of Automatic Class III Designation (De Novo). FDA.gov https://www.fda.gov/medicaldevices/deviceregulationandguidance/howtomarketyourdevice/premarketsubmissions/ucm462775.htm

Central Venous Catheters and Biofilm Infections

Bryan Haymond

Abstract Central venous catheter (CVC) use has become commonplace. Infections that have plagued mankind for centuries continue to challenge medical devices, both those that are old and those that are integrated into healthcare systems. Bacterial biofilms consist of microorganisms that create an environment with multiple characteristics including a slimy matrix and varying cell types. As these communities adhere to catheter surfaces, they can cause bloodstream infections. These are problems that surgeons and healthcare workers deal with on a daily basis. We need unique and effective therapies that can be integrated with CVC devices that will eradicate biofilms, reduce patient suffering, and improve our ability to deliver necessary medications with reduced complications.

Keywords Central venous catheter · Biofilm · Characteristics · Infection · Antimicrobial strategies

Hospital-Acquired Infections

Hospital-acquired infections (HAIs) have been plaguing us since the inception of the hospital in the twelfth century.

During this medieval time, hospitals were among the most hazardous places, with death rates as high as 70% [1]. When a sick person entered the hospital, his or her property was disposed of, and in some regions a requiem mass was held, as if the person had already died [2]. But one must remember that this was a time when ground rabbit fur and mummy powder—that is, the ground remains of mummies—were among the most popular wound dressings, and attempts at antisepsis were quite crude. Medicines of the time consisted of ingredients such as snake flesh, laurel berries, sheep dung, lye, cow kidney, antimony, alum, and earthworms that

B. Haymond (✉)
Beckton, Dickinson and Company (BD), Salt Lake City, UT, USA

D. L. Williams (ed.), *Targeting Biofilms in Translational Research, Device Development, and Industrial Sectors*,
https://doi.org/10.1007/978-3-030-30667-0_7

were mixed with various herbs and were to be taken orally or by enema [3]. Improvements to hospital infection prevention measures came very slowly over the next several centuries.

By 1800 hospital mortality was still considerable, with rates commonly above 25%. According to a report of an American military hospital at the time, thousands of young men who had been admitted to the hospital with slight injuries or venereal diseases died from serious hospital-acquired infections during their stay. It was said that "a soldier entering a great battle was in less danger than one entering the hospital." [2, 4] It was not until the Progressive Era of the late 1800s to the early 1900s that huge breakthroughs were made in the understanding of the invisible realm of microbes. During the late 1800s, the new and exciting field of bacteriology was introduced to the world, bolstered by the work of Pasteur, Koch, and Lister, and our success in combating these invisible enemies grew exponentially as a result.

Interestingly, even today, we are still fighting that age-old battle against nosocomial infections. Despite advances in medical knowledge, in pharmaceuticals, and in hospital infection prevention as a whole, we still lose patients every year to infections that are acquired while being treated for unrelated and very curable conditions, despite the fact that these infections are largely preventable. According to the CDC, around 5–10 percent of hospitalized patients in the USA are affected by HAIs. This equates to approximately 1.7 million HAIs in US hospitals every year, resulting in 99,000 deaths and an estimated $20 billion in healthcare costs [5, 6]. Of these infections, 32% are urinary tract infections (resulting from the use of urinary catheters), 22% are surgical site infections, 15% are hospital-acquired pneumonia, and 14% are bloodstream infections, primarily associated with the insertion of a vascular access device [7, 8].

Why is it that we are still losing the battle to microbes consistently? Why are we sending our loved ones to the hospital to be treated for curable diseases, only to lose them to a hospital-acquired pneumonia or bloodstream infection? For almost a century and a half we have been working under the assumption that we know and understand the invisible enemy that we are fighting. But perhaps the error lies in our fundamental understanding of the microbes themselves. Perhaps we need to accept the fact that we need to progress into the next chapter of understanding the microbial world. We need the medical community, the Food and Drug Administration (FDA), and medical device and pharmaceutical companies to get caught up with academia and advance into the paradigm of biofilms.

Biofilms

Biofilms were first explained at length by Costerton et al. in 1978. Since then biofilms have become more and more recognized and studied as a compelling field impacting medical and industrial settings alike. Microbial biofilms exist as microorganisms and produce extracellular polymers, which are used to adhere to a surface.

This extracellular polymer, known as extracellular polysaccharide (EPS), functions as a scaffolding or matrix that provides structure and security within the biofilm.

Unfortunately, in the medical community today, this is a commonly misunderstood and ignored subject. The word "biofilm" exists merely as a buzzword, and though commonly used, the meaning of the word is frequently misunderstood. Perhaps this is a branding problem. Upon hearing the word biofilm, it does conjure an image of a slimy byproduct of microorganisms, and this is precisely the root of the confusion. In an article recently published in the online publication *Science Daily*, an author attempts to introduce the subject of biofilms by saying:

> Have you ever heard of biofilms? They are slimy, glue-like membranes that are produced by microbes, like bacteria and fungi, in order to colonize surfaces. They can grow on animal and plant tissues, and even inside the human body on medical devices such as catheters, heart valves, or artificial hips. Biofilms protect microbes from the body's immune system and increase their resistance to antibiotics. They represent one of the biggest threats to patients in hospital settings.

This of course is a true statement but gives the reader the impression that the word "biofilm" only refers to the slime produced by the bacteria and not the bacteria themselves. The slime is only part of the story. Microorganisms such as bacteria exist in two main phenotypes, namely *planktonic*, which are free-floating cells, and *biofilms,* which are aggregations of cells that have adhered to a surface. Biofilms are not unique to bacteria. Fungal organisms such as yeast (*Candida* sp.) are well-known for forming biofilms. Biofilms can exist in which one species can dominate the space, known as a monomicrobial biofilm. However, biofilms rarely exist in this manner in nature; rather, they exist as polymicrobial biofilms. This is when more than one species of microorganism is well distributed throughout, at times creating a symbiotic relationship.

Biofilm development is a complex process that can be condensed into five major steps.

Stage 1—surface adherence: within minutes microorganisms can begin to colonize a surface.

Stage 2—aggregation: microcolonies form and begin to excrete EPS components, i.e., slime.

Stage 3—biofilm is formed: the community begins to mature into multilayered clusters.

Stage 4—three-dimensional growth: maturation advances to include physical pathways (water channels) that shuttle nutrients and waste products, and the biofilm begins to be protected from host defense mechanisms and antibiotics.

Stage 5—critical mass is reached: planktonic cells can escape the community and colonize other surfaces.

An oxygen gradient can also exist. Organisms such as *Staphylococcus aureus* and *Pseudomonas aeruginosa* are well-known biofilm-forming organisms and are also known collectively as facultative anaerobes. The ability to shift their physiology from that of an aerobic state where oxygen is available to an anaerobic state where oxygen is less available is advantageous. It is in these regions of limited oxygen that you will find cells that have a much lower metabolic rate and live in a state of dormancy. These cells have been referred to as persisters [9]. This ability is problematic for antibiotics whose mechanism of action is at the ribosome, for example, as the persister cells are not metabolically active; therefore, the drug will have difficulty gaining access into the bacterial cell through its usual metabolic channel and is rendered ineffective. This phenomenon of persister cells has explained the tolerance that biofilms demonstrate to multiple classes of antibiotics that require getting inside the cell to carry out their mechanism of action. Biofilms have been found to be 1000 times more tolerant to antibiotics than their planktonic counterparts, which may be contributed by persisters [10]. Cell-to-cell communication, referred to as quorum sensing, also contributes to the pathogenicity of biofilms. Quorum sensing involves signals known as autoinducers that respond to cell density and other stresses experienced in the environment, which can contribute to expression of virulence factors.

Biofilms and Bloodstream Catheters

Of all the medical procedures in the hospital setting, few are more common and ubiquitous than the insertion of a vascular access device. The establishment of reliable venous access is required for nearly every patient in the hospital regardless of their healthcare needs, and often a device is placed in a patient upon entering the emergency department whether they need it or not. Consequently, the insertion of an intravenous (IV) catheter is often the first procedure performed upon entering the hospital, and removal of an IV catheter is commonly the last.

The peripheral IV is by far the most common vascular access device utilized in hospitals around the world and is considered indispensable in modern-day medical practice [8]. These devices consist of a small flexible tube that is inserted into a peripheral vein for intravenous therapies, such as the administration of medications and fluids, and are also often used for blood draws. Today, up to 90% of patients admitted to the hospital receive a peripheral IV, and over 1 billion peripheral IVs are placed globally each year [11].

For emergent and critically ill patients, obtaining vascular access is a critical and time-sensitive management step [12]. Prompt vascular access for these patients allows for rapid laboratory testing and the administration of life-saving therapies. Often the vascular access needs for these patients exceeds the capability of a small peripheral IV catheter. For this reason, there are a variety of vascular access devices that vary in size, utility, and invasiveness to the patient. All of them, however, pose a risk of infection, due to the fact that they are percutaneous devices. As such, a risk for contamination exists from the time they are inserted to the moment they are removed.

Although catheter-related blood stream infection (CRBSI) is the subject of extensive surveillance and research, most of these efforts have been limited to the study of central venous CRBSI, while CRBSI related to peripheral IVs (PIVs) has received much less focus. Catheter-related infection is a problem that deserves attention no matter the placement or location. Reported rate of infection related to PIVs is lower than that of central venous catheters (CVCs); however, with more than 200 million PIVs being placed in the USA each year, the number of infections related to PIVs is actually greater than that of central lines [8].

CVC insertion has become an indispensable procedure in a variety of situations throughout the hospital and in home health settings. During the past half century, the multiple technical and technological achievements leading to the development of safe, short-term, long-term, or chronic vascular access have had significant effects in saving or prolonging the lives of countless patients. The many applications for CVCs include fluid resuscitation, hemodynamic monitoring, parenteral nutritional support, dialysis, and the administration of chemotherapy or other caustic or harmful medications that can't be administered peripherally.

Generally, central lines are of two main types. The first is tunneled catheters, which are implanted surgically by creating a subcutaneous track prior to entering a central vein, such as the internal jugular, subclavian, or femoral vein for long-term (weeks to months) access. These types of catheters are designed for chronic use and the indications of use including therapies such as chemotherapy and hemodialysis. The second type of central line is a "nontunneled" or acute central line. These catheters are inserted percutaneously, are the most common type of central line, and account for the majority of central-line-associated bloodstream infections (CLABSIs) that are reported [13].

Acute central line placements have long been regarded as dangerous procedures by practitioners, catheter manufacturers, and the FDA [14]. More than three million CVCs are placed annually. Of those procedures it has been reported that 3–25% experience complications [15, 16]. Common complications include inadvertent arterial injury, air embolism, pneumothorax, and CLABSI.

A CLABSI is defined as a laboratory-confirmed bloodstream infection not related to an infection at another site that develops within 48 hours of central line placement. Of all the healthcare-associated infections, CLABSIs are the most costly, accounting for approximately $46,000 per case [13]. CLABSIs lead to prolonged hospital stays and increased mortality rates. Nosocomial bloodstream infections are reported to be the eighth-leading cause of death in the USA [17]. It is estimated that more than 250,000 cases occur annually in the USA alone, with a fatality rate of approximately 23.8% [18, 19]. These incidents are costly, deadly, and largely preventable. The US Department of Health and Human Services' Action Plan to Prevent Healthcare-Associated Infections is focusing attention on the need to dramatically reduce these infections [20, 21].

Starting in 2008, CLABSIs were classified as a "never event," forcing hospitals to track and document all incidents. This increased awareness of the issue and made hospital infection rates public knowledge, increasing their incentive to address this issue and eliminate CLABSI. This has recently been reinforced by the introduction of the Affordable Care Act, which has brought about even greater awareness and monitoring.

The prevention of CLABSI is an extremely challenging and complicated issue. In 2011 the CDC took this challenge head-on when they published the "CDC

Guidelines for the Prevention of Intravascular Catheter-Related Infections." This document is comprised of 83 pages of evidence-based guidelines and instructions for the proper care and maintenance of these devices and serves as standard for healthcare personnel who insert intravascular catheters, those who are responsible for using and maintaining them, and those who are liable for the surveillance and control of infections in the hospital (infection preventionists) [22].

Although it is extremely difficult to track and confirm the source of a central line infection, there are some general perceptions about the most common causes. The incidence of catheter-related infection is directly influenced by duration of catheter dwell time in the patient. Longer dwell times result in an increased number of manipulations at the catheter hub which, in turn, can lead to increased risk of intraluminal contamination. As previously mentioned, if an infection develops within 48 hours of catheter placement, it is commonly perceived that this infection was the result of contamination during the insertion procedure. Central line insertions are sterile procedures. Much like a surgical procedure, during a central line insertion patients are draped from head to foot in a sterile barrier. The insertion site is prepared with a surgical antiseptic, typically chlorhexidine gluconate (CHG). The most common insertion sites include the internal jugular vein, the subclavian vein, a deep vessel in the upper arm, or the femoral vein. The clinician dons a sterile gown, mask, cap, and sterile gloves. Similar to other sterile procedures, the opportunity to introduce contamination is only as good as the sterile technique of the clinician performing the procedure.

It is commonly understood that within 7–10 days of CVC placement, bacteria on the surface of the skin can migrate along the surface of the catheter from the catheter insertion site towards the intravascular space. For nontunneled devices, the absence of a tunnel places these catheters at higher risk for CLABSIs. Research shows that CLABSIs that occur beyond 10 days are typically the result of contamination of the intraluminal portion of the catheter hub, and this is commonly caused by a healthcare provider's contaminated hands, often due to a breach of standard aseptic procedure while accessing the catheter. Less common mechanisms of contamination include hematogenous seeding of bacteria from another source or from a contaminated infusate [23, 24].Host factors that increase the risk of CLABSI include chronic illnesses (hemodialysis, malignancy, gastrointestinal tract disorders, pulmonary hypertension), immune-compromised states (bone marrow transplant, end-stage renal disease, diabetes mellitus), malnutrition, total parenteral nutrition (TPN), extremes of age, loss of skin integrity (burns), prolonged hospitalization before line insertion, catheter type, catheter location (femoral line has the highest, followed by internal jugular, then subclavian), conditions of insertion (emergent versus elective, use of maximal barrier precautions versus limited), catheter site care, and skill of the catheter inserter. *Pseudomonas* is commonly seen in association with neutropenia, severe illness, or known prior colonization. *Candida* is associated with other risk factors, namely femoral catheterization, TPN, prolonged administration of broad-spectrum antibiotics, hematologic malignancy, or solid organ or hematopoietic stem cell transplantation. Certain bacteria such as staphylococci, *Pseudomonas*, and *Candida* produce biofilms, which favor increased virulence, adherence to catheter surfaces, and diffidence to antimicrobial therapy [23].

Antimicrobial Strategies and Catheters

Antimicrobial-coated or impregnated central catheters were first introduced to clinical practice circa 1990 and quickly grew in popularity and clinical use in the acute setting. The two most common catheter coatings are comprised of either chlorhexidine and silver sulfadiazine, or minocycline and rifampin. For approximately six decades, chlorhexidine has been used in clinical practice as a skin antiseptic and disinfectant for a number of sterile procedures [25]. These technologies have remained unchanged in almost 30 years. Imagine how much technology has changed since that time. Antimicrobial catheters were introduced 17 years before the first iPhone. Currently, more than 75% of the acute central lines placed in the United States utilize these same antiquated coatings. It is challenging to assess the efficacy of these technologies. Since their introduction, several studies have been published evaluating their ability to reduce the incidence of CLABSI, many touting extremely positive results. In an effort to answer this question, McConnell and colleagues published a paper in 2003 critically analyzing 11 of these studies. They assessed study methodology, patient characteristics, and the presence of flaws in the studies and found that many of the studies contained inconsistent definitions of CRBI, failed to account for confounding variables, contained suboptimal statistical analyses, and lacked clinically relevant endpoints [26]. In the end, the authors concluded that although the use of impregnated catheters may decrease catheter colonization, they recommended that more reliable studies should be conducted in order to definitively conclude whether these technologies have the ability to decrease the incidence of catheter-related infection [26]. But whether they have the ability to decrease the incidence of infection or not, the problem of bloodstream infections is apparent. Antimicrobial catheter coatings have been in use for almost 30 years—why are we not doing a better job of preventing this avoidable issue?

But perhaps the question is not whether the idea of coating a catheter with antimicrobials is a valid one, but what assumptions were made about microbes in the creation and optimization of the coatings themselves. It has long been assumed by biofilm academics and enthusiasts that many of the methods established to test antimicrobial efficacy are based on a number of incorrect and outdated assumptions about bacteria themselves. From a medical device development standpoint, how can we do a better job of designing antimicrobial technologies that address the actual clinical scenario with a more complete understanding of how microbes function? As mentioned previously in this chapter, perhaps the fault lies with those that first coined the term biofilm. It leads the reader to believe that the term refers to something that microbial life creates. In reality, knowledge of biofilm is true knowledge of how microbes actually *behave* and what microbes truly are.

Knowledge is of no value unless we use it for change. Many of us who work in the medical device field do so because we believe we can make a difference in the lives of patients by elevating the technologies used to treat those that need it most. Throughout history, advancements in knowledge have led to advancements in technology and practice, which in turn have led to vast improvements in clinical care. Is it possible that a simple conceptual hang-up is preventing us from entering a new era of medical advancement? Is it possible that by simply viewing the microbial world through the biofilm lens, we might finally overcome the hurdles that are holding us back? It is my hope that a more complete understanding of microbial biofilms will allow us to overcome these hurdles, inspire the creation of new and exciting medical technologies, and finally guide us into a world where hospital-acquired infections are a thing of the past.

Disclosure The opinions and assertions included in this chapter are those of the author and do not reflect BD or affiliates.

References

1. Massengill, S. E. (1943). *A sketch of medicine and pharmacy and a view of its progress by the Massengill family from the fifteenth to the twentieth century*. SE Massengill: Bristol.
2. Smith, P. W., Watkins, K., & Hewlett, A. (2012). Infection control through the ages. *American Journal of Infection Control, 40*, 35–42.
3. Borgognnoni, T. (1955). *The surgery of Theodoric: ca AD 1267*. Campbell E, Colton J, translators. New York: Appleton Century Crofts.
4. Duncan, L. C. (1931). *Medical men in the American revolution, 1775e1783*. Carlisle: Medical Field Service School.
5. https://www.cdc.gov/washington/~cdcatWork/pdf/infections.pdf. Accessed on 1/5/2019.
6. Klevens, R. M., Edwards, J. R., Richards, C. L., Horan, T. C., Gaynes, R. P., Pollock, D. A., et al. (2007). Estimating health care-associated infections and deaths in US hospitals, 2002. *Public Health Reports, 122*, 160–166.
7. http://patientcarelink.org/improving-patient-care/healthcare-acquired-infections-hais/
8. Sato, et al. (2017). *BMC Infectious Diseases, 17*, 434. https://doi.org/10.1186/s12879-017-2536-0.
9. https://www.ncbi.nlm.nih.gov/pubmed/20528688
10. https://www.ncbi.nlm.nih.gov/pubmed/3923925
11. O'Grady, N. P. et al. CDC guidelines for the prevention of intravascular catheter related infections. (2011)
12. Ash, A., et al. (2016). Seldinger Technique for placement of "peripheral" IJ lines. *Western Journal of Emergency Medicine*.
13. Haddadin, Y., & Regunath, H. (2018). Central Line Associated Blood Stream Infections (CLABSI) [Updated 2018 Oct 27]. In *StatPearls [Internet]*. Treasure Island: StatPearls Publishing. https://www.ncbi.nlm.nih.gov/pubmed/28613641.
14. Bowdle, A. et al, Arterial cannulation during central line placement. Mirador Biomedical Report, (2009).
15. Millennium Research Group. US Markets for Vascular Access Devices. (2011).
16. Ezaru, C., et al. (2009). Eliminating arterial injury during central venous catheterization using manometry. *IARS Journal, 109*(1), 130–134.

17. Zimlichman, E., Henderson, D., Tamir, O., et al. (2013). Health care-associated infections: A meta-analysis of costs and financial impact on the US health care system. *JAMA Internal Medicine, 173*(22), 2039–2046.
18. Loftus, K., Tilley, T., Hoffman, J., Bradburn, E., & Harvey, E. (2015). Use of Six Sigma strategies to pull the line n central line-associated bloodstream infections in a neurotrauma intensive care unit. *Journal of Trauma Nursing, 22*(2), 78–86.
19. Pittet, D., et al. (1997). Microbiological factors influencing the outcome of nosocomial bloodstream infection. *Clinical Infectious Diseases, 24*, 1068–1078.
20. Services, D. O. (2013). Action plan to prevent health care-associated Infections: Road map to elimination. Retrieved from hhs.gov
21. Agency for Healthcare Research and Quality. (2011). Eliminating CLABSI A National Imperative. *AHRQ*, 1–29.
22. CDC guidelines. (2011).
23. https://www.ncbi.nlm.nih.gov/books/NBK430891/?report=reader
24. Bell, T., & O'Grady, N. P. (2017). Prevention of Central Line-Associated Bloodstream Infections. *Infectious Disease Clinics of North America, 31*(3), 551–559. [PMC free article: PMC5666696] [PubMed: 28687213].
25. Wassil, S. K. (2007). Antimicrobial impregnated catheters in the prevention of catheter-related bloodstream infection in hospitalized patients. *Journal of Pediatric Pharmacology and Therapeutics, 12*, 77–90.
26. McConnell, S. A., Gubbins, P. O., & Anaissie, E. J. (2003). Do antimicrobial-impregnated central venous catheters prevent catheter-related bloodstream infection? *Clinical Infectious Diseases, 37*, 65–72.

A Biofilm-Based Approach to the Diagnosis and Management of Postoperative Spine Infection

Jeremy D. Shaw

Abstract Postoperative spine infections are a devastating surgical complication. Historical literature reports postoperative infection rates as high as 20%. Improved surgical techniques and the use of intrawound vancomycin powder have dropped rates in recent years. Importantly, patients who experience a postoperative spine infection have a poorer perceived outcome of their surgery even if it is ultimately successful. In an era of patient-reported outcomes (PROs) driving practice patterns and an aging population undergoing increasing rates of high complexity spine surgery, infection, often complicated by biofilms, remains a key target for quality improvement. This article outlines contemporary standard of care practices for the diagnosis and treatment of postoperative spine infection with an emphasis on emerging concepts and broadly applicable surgical techniques including methylene blue staining as a disclosing agent to identify biofilm-burdened regions.

Keywords Spine surgery · Infection · Deep spine infection · Osteomyelitis · Diskitis · Biofilm

Rate of Postoperative Infection

The rate of postoperative infection remains difficult to determine due to the diverse and heterogeneous nature of spine procedures; however, the trend is clear that higher rates occur with increasing complexity, length of surgery, and invasiveness of the procedure [15, 16]. The use or absence of instrumentation appears to be a driver of infection with instrumented cases having higher rates of infection. While numbers vary per report, working numbers with which to counsel patients remain at approximately 1–2% for uninstrumented cases and approximately 5% for instrumented fusions based on prospective data [17–19]. Recent pooled average data is approxi-

J. D. Shaw (✉)
Department of Orthopaedic Surgery, University of Pittsburgh School of Medicine, Pittsburgh, PA, USA

D. L. Williams (ed.), *Targeting Biofilms in Translational Research, Device Development, and Industrial Sectors*,
https://doi.org/10.1007/978-3-030-30667-0_8

mately 1.9% for all spine cases [20]. For thoracolumbar deformity cases, self-reported Scoliosis Research Society data reports an overall infection rate of 2.1% while more recent International Spine Study Group (ISSG) reports a 2.4% rate of deep infection [5, 21]. These numbers, however, must be interpreted with caution due to inherent bias and systemic underreporting of infection data [22].

Clearly defined modifiable and nonmodifiable risk factors for postoperative spine infection are well documented. Spinal trauma patients represent a unique population that have an increased risk for developing postoperative infections. The elevated infectious risk for this population is primarily attributed to damage to the soft tissue envelope leading to local tissue hypoxia with subsequent necrosis, edema, acidosis and hematoma, thus creating the ideal milieu for bacterial proliferation [23]. Trauma patients also are in a state of systemic paradoxical immunosuppression from the traumatic event, which is further thought to increase susceptibility to infection [24]. Comorbid factors such as age, nutritional status, body habitus, and other medical conditions cannot be controlled for in the same manner as they are in elective surgery and further compound infectious risk. Consequently, the rate of postoperative infection in this population is approximately two to three times higher than nontrauma cases [25–27].

Spinal surgeries for management of tumors are also associated with significantly higher rates of postoperative infections with those receiving local radiation at particular risk [28, 29]. It is generally recommended that patients not undergo surgery within 6–12 weeks of preoperative radiation or receive postoperative radiation within 3 weeks of surgery in order to allow adequate soft tissue healing [17].

Nonmodifiable risk factors must be evaluated and maximally treated prior to surgery. These include conditions such as rheumatoid arthritis, human immunodeficiency virus/acquired immunodeficiency syndrome (HIV / AIDS), psychiatric illness, substance abuse, and corticosteroid use all of which have been linked to elevated risk of infection. While age is not an independent risk factor for postoperative spine infection, age is correlated with increased medical comorbidity which is a known risk factor for infection [17, 30].

Modifiable risk factors include smoking, obesity, procedure length, catheter use, length of hospital stay, and malnutrition. Poorly controlled diabetics are at particular risk [30]. Anterior procedures and minimally invasive surgery (MIS) procedures appear to have correspondingly lower rates of postoperative infection in most cases, likely due to the preserved and robust soft tissue envelope left largely undisturbed [17, 22, 31–33]. In aggregate, modifiable and nonmodifiable risk factors are perhaps best summarized in the emerging concept of patient frailty, which appears positively correlated with elevated rates of postoperative infection in frail patients [34].

Definition and Diagnosis of Postoperative Spine Infection

Importantly, there are no clearly stated sets of diagnostic criteria which define a postoperative spine infection. Increased pain, fever, and wound erythema are present in less than 30% of cases. The most reliable marker seems to be increased wound

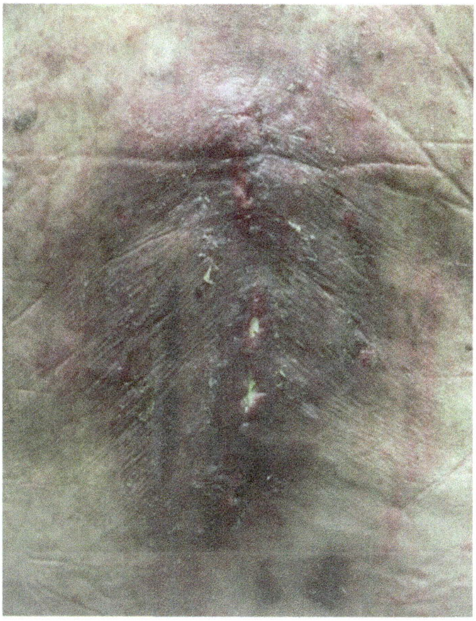

Fig. 1 Macerated dorsal spine wound in the early postoperative period with increasing drainage

drainage at 10–14 days, which occurs in two-thirds of postoperative spine infection cases (Fig. 1) [22]. For deep infection, often there is a pain-free period after surgery for 1 to 2 months and subsequently increasing pain or development of new neurologic symptoms over several weeks. Pain is often out of proportion to what would otherwise be expected. These findings are often associated with constitutional symptoms. Superficial wound infections, in contrast, typically present at 1 to 2 weeks postoperatively and are less frequently associated with constitutional symptoms. Superficial wound infections can most commonly be treated with wound care and oral antibiotics [17].

If there is concern for deep underlying infection, additional work-up is warranted. Laboratory values are the first line of additional diagnostics in cases of suspected postoperative infection. Initial blood work-up should consist of white blood cell count (WBC), erythrocyte sedimentation rate (ESR), and C-reactive protein (CRP). Of commonly assessed laboratory values, CRP has the highest diagnostic sensitivity for the identification of a postoperative spine infection [35, 36]. It is particularly useful for the diagnosis of postoperative spine infection as it normalizes quickly following spine surgery. In uninfected patients it should return to baseline 3–7 days postoperatively [37]. In contrast, ESR may not normalize for 3 to 6 weeks following an invasive procedure, decreasing diagnostic utility in the early postoperative period [38]. Less commonly assessed markers such as procalcitonin, serum amyloid-A protein, and interleukin-6 (IL-6) have also been evaluated in the literature for the diagnosis of postoperative infection and have been found to have high sensitivity and to be superior to CRP in several studies [39–41].

The accurate identification of the infectious organism is a critical step in the treatment of postoperative spine infection. While some authors advocate superficial wound cultures, these experience high rates of contamination with local skin flora and can complicate the diagnostic work-up. If there is a fluid collection, early aspiration, however, may be beneficial for diagnosis [42]. Computed tomography (CT) or fluoroscopic guidance may be used to obtain a fine-needle aspiration or preferably a core biopsy of the affected area [17]. The most accurate cultures, however, are those obtained during surgery. Unfortunately, even when intraoperative cultures are obtained at the time of surgery, they are often negative in patients with established postoperative spine infections. The diagnostic sensitivity of cultures is further worsened since many patients receive antibiotics prior to obtaining intraoperative cultures [22].

Tissue cultures remain the gold standard for infection diagnosis in spine surgery [17]. However, other subspecialty domains, particularly arthroplasty, have embraced novel molecular biology techniques which have proven particularly useful in identification of culture-negative infections. These techniques can often identify infection even in presumptively aseptic revision settings. Implant sonication, polymerase chain reaction (PCR), and next-generation sequencing are available, if underutilized, diagnostic techniques with broad applicability to spine surgery [43–46].

Indeed, the current state of diagnosis for postoperative spine infection is poorly defined. This lies in contrast to the arthroplasty literature which has defined and frequently updated consensus-based diagnostic criteria for infection of a prosthetic joint. The initial definitions for periprosthetic joint infection (PJI) were published in 2011 and have been subsequently updated and validated in 2013 and 2018, respectively [22, 47–49].

Unfortunately, no similar consensus definition can be applied to the arena of spinal surgery. However, recently updated guidelines for the diagnosis of periprosthetic joint infection provide an excellent starting point to define postoperative or periprosthetic spine infection. Specifically, patients with a sinus tract communicating to the hardware or bone or those with two positive cultures of the same organism can likely be presumed infected. Similarly, those with an intraoperative constellation of positive histology, purulence, and/or a single positive culture can likely be presumed infected. These findings, however, do not necessarily help with the decision of whether or not to return to the operating room to treat a presumed infection. In that regard, elevated serum CRP, D-dimer, and erythrocyte sedimentation rate (ESR) may be most helpful and are commonly assessed in the setting of infection. To the authors knowledge, analysis of local fluid white blood cell count, leukocyte esterase, alpha-defensin, polymorphonuclear (PMN) cell percentage, and CRP have not been evaluated in the setting of postoperative spine infection; however, these markers may provide diagnostic value based on extrapolation of current arthroplasty literature [48].

Imaging

Plain film radiographs are the first imaging that should be obtained as part of a diagnostic work-up for suspected infection. It may take up to 4 weeks for radiographs to show evidence of infection; however, subtle bony lysis at the bone-prosthetic interface and implant loosening are early clues. Infectious disk space changes may take longer to develop and are often challenging to differentiate from degenerative changes. More substantial bony changes such as osteolysis, end plate destruction, and deformity typically take 2 months or more. Paravertebral soft tissue swelling is also a strong indicator of potential abscess, particularly in the retropharyngeal space or paraspinal musculature [17, 50].

CT provides a more detailed view of bony anatomy and allows for earlier detection of infection-related bony changes when compared to plain radiographs. When IV contrast is used, CT can also provide clues to soft tissue collections that are not identifiable on plain radiographs and can be useful in patients who are not candidates for magnetic resonance imaging (MRI). While nuclear imaging modalities such as gallium, technetium, and indium bone scan have been demonstrated to have limited utility, positron emission tomography (PET) and PET-CT have an emerging role in the diagnosis of postoperative spine infections that may have otherwise equivocal imaging [17, 51–53].

MRI with and without contrast remains the gold standard used for clinical decision making in the setting of postoperative spine infection. For the diagnosis of postoperative spine infection, it is both highly sensitive and highly specific; however, as with other modalities it can be difficult to distinguish early nonpathologic postoperative changes from infections [17, 54–56]. Of particular utility may be the recently described pedicle screw sign which is defined as fluid collection outside the head of the pedicle screw, which was represented by a high intensity area extending more than 5 mm outside the lateral edge of the head of the screw in the T2-weighted axial plane (Fig. 2) [57]. A metal artifact, particularly with stainless steel or cobalt, can further limit the diagnostic utility of MRI [54–58].

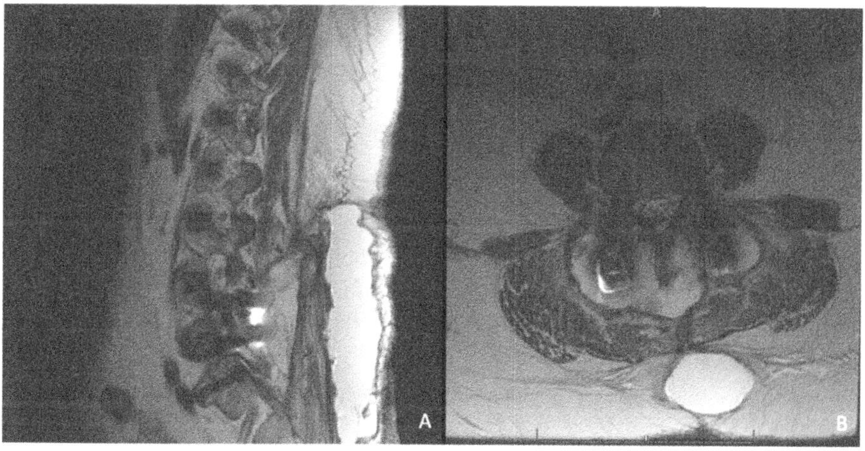

Fig. 2 (**a**) Parasagittal T2-weighted MRI showing superficial and deep fluid collections. (**b**) The pedicle screw sign can be seen with fluid collections extending more than 5 mm outside the lateral edge of the head of the pedicle screw in the axial plane. This wound should be presumed infected unless proven otherwise

Microbiology

Generally, three mechanisms are described for postoperative infections – direct inoculation during the procedure, contamination during the early postoperative period, and hematogenous seeding. Of these three, direct inoculation during the surgery is the most common [17].

Gram-positive cocci are the most common pathogens responsible for acute postoperative spine infections. Of these, *Staphylococcus aureus* causes more than 50% of infections in some reports with *S. epidermidis* and β-hemolytic streptococci as the next most common. Common gram-negative pathogens include *Escherichia coli*, *Pseudomonas aeruginosa*, *Klebsiella pneumoniae*, *Enterobacter cloacae*, *Bacteroides*, and *Proteus* species. The anatomic location of the wound impacts the likelihood of gram- negative infection, with lumbosacral incisions having an increased risk of gram-negative infection due to fecal and urinary contamination. Additionally, in cases of patients who are immunosuppressed, fungal infection is also a risk [17, 59, 60].

While vancomycin powder has reduced the overall rate of infection following spine surgery [1–4, 8–14], there is a growing body of evidence that shows the traditional microbial profile of postoperative spine infections to be changing. Due in large part to the use of in-wound vancomycin powder and associated killing of gram-positive organisms, there is significant selection pressure for gram-negative organisms. Thus, while the overall number of infections is greatly decreased, the proportion of gram-negative infections has increased [22, 61, 62].

For late spine infections which present a year or more after spine surgery, low virulence organisms such as *Cutibacterium acnes* (formerly *Propionibacterium acnes*) are the most common causative agents. These organisms are postulated to be present in normal skin flora and contaminate the wound via intraoperative inoculation or prolonged drainage and inflammation. Importantly, if *C. acnes* is suspected as an infectious agent, cultures need to be retained by the microbiology lab for 2 weeks [63–65].

With a delayed onset of presentation, hematogenous spread of infection must also be considered. These infections are typically due to highly virulent organisms and often present in patients with systemic illness, intravenous drug use, immunosuppression, and sepsis [66].

Most periprosthetic infections, including postoperative spine infections, are caused by biofilm-forming organisms [67]. Basic science and animal literature suggest that biofilms are established in vivo within hours to days [68–72]. Importantly, for periprosthetic infections caused by biofilm-forming organisms, there is no literature to support the antithetical position that there is clinically significant periprosthetic infection without biofilm.

Prevention

The easiest way to manage postoperative spine infection is prevention. Hospital and medical system factors play a role in the rate of postoperative infection. Preoperative nasal methicillin-resistant *S. aureus* (MRSA) colonization is associated with postoperative spinal MRSA. Preoperative screening and subsequent decolonization using topical antibiotics have been shown to reduce the rate of surgical site infection and are cost-effective [73, 74]. Case order and seasonality also impact the rate of surgical site infection after spine surgery with cases occurring later in the day having higher rates of infection, as well as those during the summer months [75, 76]. When using implants in spine surgery, keeping the instrumentation covered if opened at the beginning of the case, or not opening until necessary, leads to lower colonization rates which may lead to lower infection [77].

Decision-Making

Infection prevention starts with good patient selection. Obese patients are considered at high risk for developing postoperative infections. Specifically, it appears that the distribution of body mass actually is even more predictive of surgical site infection (SSI) than absolute body mass index (BMI), with MRI measurements of skin-to-lamina distance and thickness of subcutaneous adipose layer being significant risk factors [78]. Those with an excessively thick layer of subcutaneous fat are at an elevated risk of postoperative spine infection and should be counselled accordingly. Surgery should not necessarily be delayed or cancelled, as obese patients have a treatment effect associated with surgery that is at least equivalent to nonobese individuals. This is in large part due to inferior outcomes with nonoperative management in obese patients [79]. Additionally, new studies suggest that bariatric surgery before elective posterior lumbar fusion may mitigate risk of medical complications and postoperative spine infection [80].

While diabetic patients have a higher risk of postoperative spine infection vs their nondiabetic counterparts, all diabetics are not the same. Insulin-dependent diabetic patients have a different risk profile vs noninsulin-dependent diabetics, with those requiring insulin experiencing both more and more severe perioperative complications, including infection [81]. Similarly, elevated preoperative hemoglobin A1c (HbA1c) has been linked to an elevated infectious risk, with patients having a HbA1c >7.0% at an elevated risk [82].

Recent studies have shown that both cervical and lumbar spine surgery within 3–6 months following epidural steroid injection may be associated with an increased rate of postoperative infection. Thus increasing the time interval between injection and spine surgery to at least 3 or possibly 6 months may decrease infection rates [83, 84].

Intraoperative Measures

For preoperative surgical skin antisepsis, spine surgeons continue to use both iodine- and chlorhexidine-based agents. While a small prospective series examined both chlorhexidine- and iodine-based agents and found no difference in antiseptic properties in the lumbar spine, broader literature suggests the likely superiority of alcohol-based agents, specifically chlorhexidine-isopropyl alcohol [85–87].

Preoperative weight-based antibiotic prophylaxis within 60 minutes prior to incision remains the standard of care for spine surgery with demonstrated benefit in reduction of postoperative infection [88]. Cefazolin is the antibiotic of choice, with clindamycin and vancomycin as acceptable options if cefazolin is not possible due to contraindication [87, 89]. There is currently no role for routine use of vancomycin alone. In patients known to be colonized with MRSA or at risk for MRSA colonization such as patients with recent hospitalization, nursing home residents and those on hemodialysis vancomycin may be used in addition to cefazolin. Dual coverage is preferred as vancomycin is less effective than cefazolin for preventing surgical site infections caused by methicillin-sensitive *Staphylococcus aureus* (MSSA) [89–93].

Antibiotic-containing irrigation has long been used across spine and multiple surgical domains; the literature is mixed on their performance, as well as possible effects on bone and soft tissue healing as well as if high- or low-pressure systems are preferred. Use of these agents in in vitro studies demonstrate reduced bacterial counts; however, there are no significant trials that clearly support the use of antibiotic irrigation in spinal surgery [17]. There is, however, mounting evidence that irrigation with dilute betadine may be beneficial prior to wound closure [94, 95]. Additionally, betadine appears less toxic than other antimicrobial wound cleansers [96–98].

Application of in-wound antibiotic has been popularized by the marked reductions achieved in postoperative infection rates across a variety of procedures in both adult and pediatric populations [6, 7, 99–104]. Antibiotics, most commonly vancomycin, are placed in the wound prior to closure at the conclusion of the case. They may also be mixed with the bone graft in the case of fusion type procedures. Importantly this does not appear to inhibit bony fusion [105]. Tobramycin and gentamicin are also popular options with enhanced gram-negative bacterial coverage [106–109]. Due to morbidity and cost of postoperative spine infections, the use of these intraoperative adjuncts has proven highly cost-effective [110]. There are few known downsides; however, sterile seroma and circulatory collapse have been documented as case reports [111, 112]. Early concerns about increased topical antibiotic use causing antibiotic resistance have not born out in the literature. This is postulated to be a result of supratherapeutic levels of antibiotic causing early wound bed sterilization [113].

In contrast to other purported infection-reducing techniques, use of iodine-impregnated adhesive drapes does not appear to reduce the rate of surgical site infection [114]. Similarly, use of closed suction drainage appears to have no effect on infection rates [115]. Rather, drain use has been linked to increased transfusion rates [116]. Transfusion rates have been independently associated with increased rates of postoperative infection [117, 118]. Thus, use of surgical drains should be judicious.

Perhaps most importantly, attention to detail and basic principles of sterile technique remains essential. General operating room behavior, which may create numerous opportunities for small violations in sterile technique, has been attributed to higher rates of surgical site infection. Indeed, current evidence suggests that positive intraoperative cultures occur in nearly 1/3 of primary deformity cases [119]. Thus, common sense actions such as appropriate hand washing, frequent glove changes, covering implants while not in use, and minimizing operating room traffic all contribute to lower rates of postoperative spine infection [77, 120–122].

Management of Postoperative Spine Infections

Successful treatment of postoperative spine infection requires timely and appropriate diagnosis, as well as coordinated medical and surgical management. The goal of treatment is eradication of infection which must be accomplished while maintaining vertebral column stability. The obligate requirement of stability differentiates treatment of postoperative spine infections from other postoperative and implant-associated infections as implant removal may not be feasible.

The role of biofilm in postoperative spine infections is underappreciated. Bacterial biofilms pose a major challenge in treating periprosthetic spine infections as they provide bacteria substantial protection against antimicrobial agents and the host immune response [123]. Antibiotics are unable to effectively eradicate all phenotypes of cells within a biofilm. Specifically, a phenotypic "persister" subset of biofilm cells are tolerant of antibiotic concentrations many orders of magnitude greater than would otherwise kill planktonic phenotype. Biofilms are a nidus of infection as their persisters outlast antibiotic treatments and subsequently reseed infection [124]. Thus, surgical debridement is essential to the eradication of biofilm-associated infections. However, knowing which tissue should be removed and which should remain is highly dependent on a surgeon's experience [125, 126].

Conventional spine wisdom suggests that all dermal margins that appear infected should be excised as well as all subcutaneous tissues, including fascia, that are in contact with infectious or necrotic material. If underlying deep fascial layers appear intact, some authors advocate limited subcutaneous debridement; however, there is usually some communication between superficial and deep surgical planes and missing a deep infection is potentially disastrous [17]. While a viable bone graft may be retained, any bone graft that is in contact with infection or necrotic tissue should also be removed (Fig. 3) [17, 64].

At this time, the need for repeated surgical debridement or hardware removal and exchange is driven by surgeon preference. Some authors recommend a "second-look" irrigation and debridement at 48–72 hours after the initial debridement in all cases; however, this is not the norm in clinical spine practice [17]. To better risk stratify patients requiring repeated debridement, Dipaola et al. developed a postoperative infection treatment score for the spine (PITSS). The general message is that sick patients with highly virulent polymicrobial or MRSA infections, hardware, and allograft are at high risk of infectious failure with a single-stage irrigation and debridement [127]. While novel within the arena of spine surgery, critical analysis of this article indicates that the authors fail to appreciate the underlying reason for infectious failure is likely a residual bacterial biofilm.

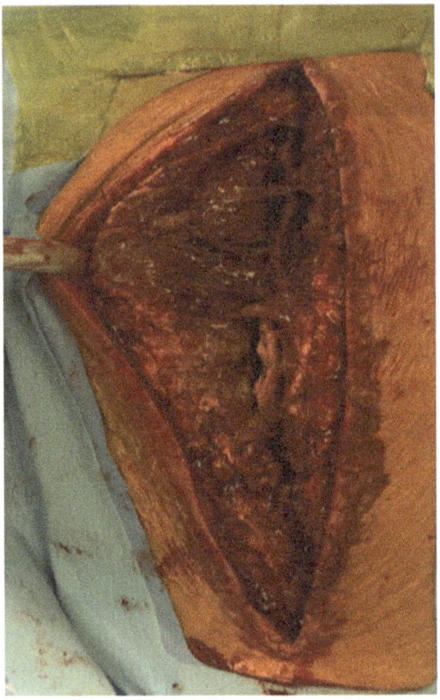

Fig. 3 Postoperative infection with gross purulence and deep necrotic muscle

This concept is best explained by examination of literature relating to irrigation and debridement for acute periprosthetic joint infection (PJI) and the debate about single- vs two-stage exchange for chronic PJI. In both clinical scenarios, the ability to eradicate tenacious bacterial biofilms appears essential to reliable eradication of deep periprosthetic infections [128, 129]. Nonetheless, both in spine and arthroplasty, the use of irrigation and debridement to treat infection likely persists because of the perceived radical option of two-stage exchange to achieve infection control. While host factors and virility of the organism play a role, the inability of parenteral antibiotics to eliminate all cell types in the biofilm layers embedded on the implant and host tissue is thought to be the primary reason for the failure of this treatment option [128].

Given that residual biofilm on both implant and host tissue is postulated to be the common mode of failure for management of deep periprosthetic infections, a technique to reliably identify biofilm in the operative settings holds promise for reducing the failure rate and consequent morbidity, mortality, and cost. Adequate debridement, however, is complicated by inability to visualize most biofilms with the naked eye. To that end, methylene blue has recently shown promise as a biofilm-disclosing agent in the orthopedic literature in both in vitro and in vivo settings and may have utility for the treatment of deep spine infections (Fig. 4) [130–133]. Indeed, preliminary data suggests that methylene blue is able to bind and stain components of *S. aureus* and *P. aeruginosa* biofilms on a variety of implant surfaces and will not stain most healthy host tissue substrates (Fig. 5).

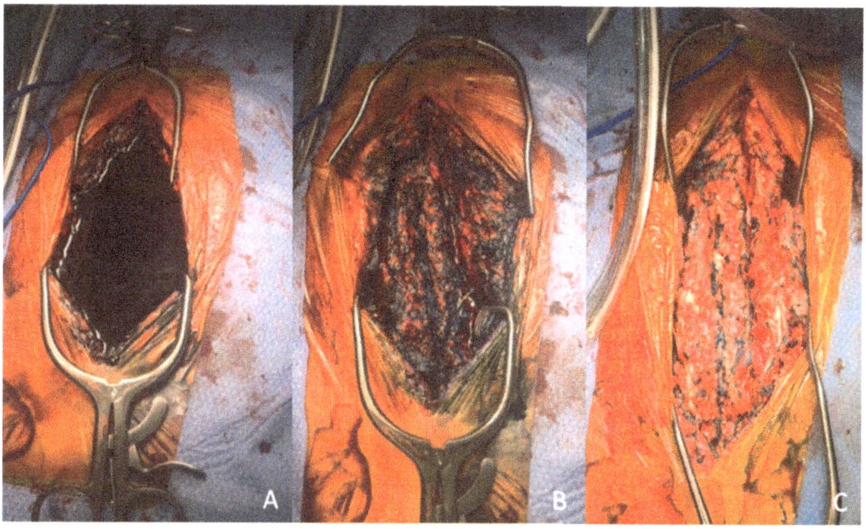

Fig. 4 Visual description of methylene blue technique. (**a**) Dilute methylene blue solution is instilled in the wound after opening the incision. (**b**) Residual dye is removed, the wound is irrigated, and the remaining blue dye stains infected and necrotic tissue. (**c**) Blue tissue is debrided leaving a healthy appearing wound bed

Fig. 5 Example of *P. aeruginosa* biofilms on polyethylene. (**a**) Biofilms of *P. aeruginosa* grown in a CDC biofilm reactor on a polyethylene coupon, stained with methylene blue and imaged by digital photography. (**b**) Scanning electron microscope (SEM) image of *P. aeruginosa* biofilms on the polyethylene surface. Circumferential ridges are machining marks (yellow arrow). (**c**) High magnification SEM image of *P. aeruginosa* biofilms on the polyethylene surface

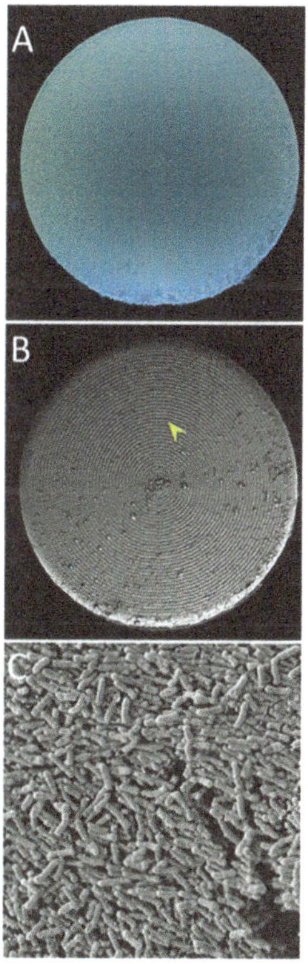

During the debridement of infections with instrumentation, implants should be inspected and replaced if there are obvious signs of loosening or failure. However, removal of infected instrumentation that remains well fixed is highly controversial. The literature on this topic is conflicted, with some authors reporting successful eradication of both anterior and posterior infections with retained instrumentation. However, a recent trend, particularly within the arena of spinal deformity surgery, is complete removal of all instrumentation independent of fixation or fusion status because of the difficulty of eliminating infection without removal [134–138]. Indeed, residual biofilm on spine implants is associated with infectious failure and need for additional surgery. In this regard, there has been a significant shift towards hardware removal if deep infection is suspected. A recent MRI-based study concluded that once vertebral osteomyelitis or intervertebral abscess was evident in MRI images, all the hardware should be removed [139]. If hardware is not able to be removed, long-term antibiotic suppression may be required until fusion is achieved and implants can be removed.

Adjunctive Surgical Techniques

Surgical techniques in addition to the application of in-wound antibiotic powder include the placement of antibiotic-containing beads. This can be done either as part of a single- or multistage surgical debridement strategy. Use of antibiotic-containing polymethylmethacrylate (PMMA) bone cement or bioabsorbable calcium sulfate beads is a viable option. These products have prolonged elution characteristics versus powdered antibiotics alone and may provide longer-term local antibiotic delivery. These products are most commonly used with vancomycin, tobramycin, and/or gentamicin as they are heat stable. Other antibiotic options are available and should be based on preoperative culture data [140–143]. For difficult-to-treat fungal infections of the spine, amphotericin B and voriconazole are both heat stable and may also be added to bone cement [144, 145].

Achieving reliable fusion following postoperative spine infection is particularly challenging. Rates of pseudarthrosis and subsequent hardware failure are elevated. This may be due in part to the ability of bacteria to impair fusion, colonization of instrumentation, and impaired vascularity in fusion beds. Consequently, even use of iliac crest bone graft (ICBG), long considered the standard of spinal fusion, cannot ensure reliable bony fusion. While the initial FDA labeling of recombinant human bone morphogenetic protein-2 (rhBMP-2, Infuse, Medtronic) listed active infection as a contraindication, several series have successfully published on the use of BMP to successfully achieve bony fusion in the setting of difficult-to-treat infection [146–148]. While more research on this topic is needed, this may be a useful adjunct to achieve fusion in an inhospitable host environment.

Severe postoperative spinal infections may result in significant soft tissue defects that require complex wound management. Early involvement of plastic and reconstructive surgeons is essential in optimizing patient outcome in these settings.

Plastic surgeons should be involved prior to definitive spine management. Ultimately these complex wounds may require flap coverage or healing by secondary intention. In both regards, vacuum-assisted closure (VAC) devices have been used successfully. VAC technology is particularly helpful in closing complex wounds, as the application of negative pressure assists in the development of granulation tissue, promotes angiogenesis, increases responsiveness to growth factors, and decreases bacterial levels. Recent literature also suggests that they may safely be placed directly on the dura even if there is no intervening soft tissue [149–152]. Local, rotational, and free muscle and tissue flaps may also be used to bring increased vascularity and adequate soft tissue coverage while protecting instrumentation and allowing bony fusion [153, 154]. Trapezius muscle flaps have historically been the gold standard for cervical and thoracic coverage; however, paraspinous muscle flaps have also gained in popularity [155, 156].

Medical Management

Culture-based parenteral antibiotic therapy remains a mainstay of treatment for postoperative spine infections. Currently, these are treated with a minimum of 6 weeks and possibly 3 months of intravenous antibiotics, followed by additional oral antibiotics. Oral regimens often include rifampin, which is thought to be beneficial in the treatment of biofilm-forming organisms [157, 158]. Difficult-to-treat or recurrent infection may also require longer-term or even lifetime antibiotic suppression [159]. Postoperative diskitis and epidural abscess are typically treated initially with antibiotic regimens unless surgery is indicated for neurologic compromise or recalcitrant progressive infection [17]. As knowledge of biofilm-based periprosthetic infections improves, more evolved approaches to antibiotic therapy will likely become available. Anticipated advances in this arena include both novel agents and innovative combinations of existing drugs that together have improved ability to target different bacterial subpopulations in difficult-to-treat biofilm-based infections.

Conclusions

Spine infection rates likely range from 1% to 5% percent based on prospective data. Recent retrospective data puts aggregate rates for postoperative spine infection at approximately 2%. Rates vary by procedure and increase with surgical invasiveness. Modifiable and nonmodifiable risk factors should be maximally managed prior to surgery, and attention should be paid to patient frailty. CRP represents the single best laboratory value to follow in the setting of postoperative spine infection. MRI with and without gadolinium contrast remains the imaging modality of choice to supplement plain film radiographs in the diagnosis of infection. Accurate and timely diagnosis of infectious organism is crucial to long-term infection eradication and disease-free survival. Novel molecular biological techniques such as PCR and next-generation sequencing should be considered in the setting of culture-negative infection and suspicious aseptic revision surgery. Appropriate antibiotic therapy remains essential. Surgical debridement remains a mainstay in the treatment of postoperative spine infections and is essential for eradication of biofilm-associated infections with or without implant retention.

References

1. El-Gindi, S., Aref, S., Salama, M., & Andrew, J. (1976). Infection of intervertebral discs after operation. *Journal of Bone and Joint Surgery, 58*, 114.
2. Pilgaard, S. (1969). Discitis (closed space infection) following removal of lumbar intervertebral disc. *The Journal of Bone and Joint Surgery*. American Volume, *51*, 713. https://doi.org/10.2106/00004623-196951040-00009.
3. Lindholm, T. S., & Pylkkänen, P. (1982). Discitis following removal of intervertebral disc. *Spine (Phila Pa 1976), 7*, 618. https://doi.org/10.1097/00007632-198211000-00018.
4. Stolke, D., Sollmann, W. P., & Seifert, V. (1989). Intra- and postoperative complications in lumbar disc surgery. *Spine (Phila Pa 1976), 14*, 56. https://doi.org/10.1097/00007632-198901000-00011.
5. Smith, J. S., Shaffrey, C. I., Sansur, C. A., Berven, S. H., Fu, K. M. G., Broadstone, P. A., et al. (2011). Rates of infection after spine surgery based on 108,419 procedures: A report from the Scoliosis Research Society morbidity and mortality committee. *Spine (Phila Pa 1976), 36*, 556. https://doi.org/10.1097/BRS.0b013e3181eadd41.
6. Heller, A., McIff, T. E., Lai, S. M., & Burton, D. C. (2015). Intrawound vancomycin powder decreases staphylococcal surgical site infections after posterior instrumented spinal arthrodesis. *Journal of Spinal Disorders & Techniques, 28*, E584. https://doi.org/10.1097/BSD.0000000000000045.
7. Chiang, H. Y., Herwaldt, L. A., Blevins, A. E., Cho, E., & Schweizer, M. L. (2014). Effectiveness of local vancomycin powder to decrease surgical site infections: A meta-analysis. *The Spine Journal, 14*, 397. https://doi.org/10.1016/j.spinee.2013.10.012.
8. Caroom, C., Tullar, J. M., Benton, E. G., Jones, J. R., & Chaput, C. D. (2013). Intrawound vancomycin powder reduces surgical site infections in posterior cervical fusion. *Spine (Phila Pa 1976), 38*, 1183. https://doi.org/10.1097/BRS.0b013e31828fcfb5.
9. Evaniew, N., Khan, M., Drew, B., Peterson, D., Bhandari, M., & Ghert, M. (2015). Intrawound vancomycin to prevent infections after spine surgery: A systematic review and meta-analysis. *European Spine Journal, 24*, 533. https://doi.org/10.1007/s00586-014-3357-0.
10. Khan, N. R., Thompson, C. J., DeCuypere, M., Angotti, J. M., Kalobwe, E., Muhlbauer, M. S., et al. (2014). A meta-analysis of spinal surgical site infection and vancomycin powder. *Journal of Neurosurgery. Spine, 21*, 974. https://doi.org/10.3171/2014.8.SPINE1445.
11. Sing, D. C., Khanna, R., Shaw, J. D., Metz, L. N., Burch, S., & Berven, S. H. (2016). Increasing rates of surgical management of multilevel spinal curvature in elderly patients. *Spine Deformity, 4*, 365–372. https://doi.org/10.1016/j.jspd.2016.03.005.
12. Hart, R. A., Cabalo, A., Bess, S., Akbarnia, B. A., Boachie-Adjei, O., Burton, D., et al. (2013). Comparison of patient and surgeon perceptions of adverse events after adult spinal deformity surgery. *Spine (Phila Pa 1976), 38*, 732. https://doi.org/10.1097/BRS.0b013e31827ae242.
13. Petilon, J. M., Glassman, S. D., Dimar, J. R., & Carreon, L. Y. (2012). Clinical outcomes after lumbar fusion complicated by deep wound infection: A case-control study. *Spine (Phila Pa 1976), 37*, 1370. https://doi.org/10.1097/BRS.0b013e31824a4d93.

14. Falavigna, A., Righesso, O., Traynelis, V. C., Teles, A. R., & da Silva, P. G. (2011). Effect of deep wound infection following lumbar arthrodesis for degenerative disc disease on long-term outcome: A prospective study: Clinical article. *Journal of Neurosurgery Spine, 15*, 399. https://doi.org/10.3171/2011.5.SPINE10825.
15. Mirza, S. K., Deyo, R. A., Heagerty, P. J., Konodi, M. A., Lee, L. A., Turner, J. A., et al. (2008). Development of an index to characterize the "invasiveness" of spine surgery: Validation by comparison to blood loss and operative time. *Spine (Phila Pa 1976), 33*, 2651. https://doi.org/10.1097/BRS.0b013e31818dad07.
16. Cizik, A. M., Lee, M. J., Martin, B. I., Bransford, R. J., Bellabarba, C., Chapman, J. R., et al. (2012). Using the spine surgical invasiveness index to identify risk of surgical site infection: A multivariate analysis. *The Journal of Bone and Joint Surgery. American Volume, 94*, 335–342. https://doi.org/10.2106/JBJS.J.01084.
17. Garfin, S., Eismont, F., Bell, G., & Bono, C. F. J. (2017). *Rothman-Simeone and Herkowitz's The Spine*. Philadelphia: Elsevier.
18. Weinstein, J. N., Tosteson, T. D., Lurie, J. D., Tosteson, A., Blood, E., Herkowitz, H., et al. (2010). Surgical versus nonoperative treatment for lumbar spinal stenosis four-year results of the spine patient outcomes research trial. *Spine (Phila Pa 1976), 35*, 1329. https://doi.org/10.1097/BRS.0b013e3181e0f04d.
19. Weinstein, J. N., Tosteson, T. D., Lurie, J. D., Tosteson, A. N. A., Hanscom, B., Skinner, J. S., et al. (2006). Surgical vs nonoperative treatment for lumbar disk herniation: the Spine Patient Outcomes Research Trial (SPORT): a randomized trial. *JAMA, 296*, 2441. https://doi.org/10.1001/jama.296.20.2441.
20. Patel, H., Khoury, H., Girgenti, D., Welner, S., & Yu, H. (2017). Burden of surgical site infections associated with select spine operations and involvement of *Staphylococcus aureus*. *Surgical Infections, 18*, 461. https://doi.org/10.1089/sur.2016.186.
21. Sciubba, D. M., Yurter, A., Smith, J. S., Kelly, M. P., Scheer, J. K., Goodwin, C. R., et al. (2015). A comprehensive review of complication rates after surgery for adult deformity: A reference for informed consent. *Spine Deformity, 3*, 575. https://doi.org/10.1016/j.jspd.2015.04.005.
22. Radcliff, K. E., Neusner, A. D., Millhouse, P. W., Harrop, J. D., Kepler, C. K., Rasouli, M. R., et al. (2015). What is new in the diagnosis and prevention of spine surgical site infections. *The Spine Journal, 15*, 336. https://doi.org/10.1016/j.spinee.2014.09.022.
23. Kälicke, T., Schlegel, U., Printzen, G., Schneider, E., Muhr, G., & Arens, S. (2003). Influence of a standardized closed soft tissue trauma on resistance to local infection. An experimental study in rats. *Journal of Orthopaedic Research, 21*, 373. https://doi.org/10.1016/S0736-0266(02)00149-3.
24. Gentile, L. F., Cuenca, A. G., Efron, P. A., Ang, D., Bihorac, A., McKinley, B. A., et al. (2012). Persistent inflammation and immunosuppression: A common syndrome and new horizon for surgical intensive care. *Journal of Trauma and Acute Care Surgery, 72*, 1491. https://doi.org/10.1097/TA.0b013e318256e000.
25. Blam, O. G., Vaccaro, A. R., Vanichkachorn, J. S., Albert, T. J., Hilibrand, A. S., Minnich, J. M., et al. (2003). Risk factors for surgical site infection in the patient with spinal injury. *Spine (Phila Pa 1976), 28*, 1475. https://doi.org/10.1097/01.BRS.0000067109.23914.0A.
26. Rechtine, G. R., Bono, P. L., Cahill, D., Bolesta, M. J., & Chrin, A. M. (2001). Postoperative wound infection after instrumentation of thoracic and lumbar fractures. *Journal of Orthopaedic Trauma, 15*, 566. https://doi.org/10.1097/00005131-200111000-00006.
27. Ghobrial, G. M., Maulucci, C. M., Maltenfort, M., Dalyai, R. T., Vaccaro, A. R., Fehlings, M. G., et al. (2014). Operative and nonoperative adverse events in the management of traumatic fractures of the thoracolumbar spine: a systematic review. *Neurosurgical Focus, 37*, E8. https://doi.org/10.3171/2014.4.FOCUS1467.
28. Omeis, I. A., Dhir, M., Sciubba, D. M., Gottfried, O. N., McGirt, M. J., Attenello, F. J., et al. (2011). Postoperative surgical site infections in patients undergoing spinal tumor surgery: Incidence and risk factors. *Spine (Phila Pa 1976), 36*, 1410. https://doi.org/10.1097/BRS.0b013e3181f48fa9.

29. McPhee, I. B., Williams, R. P., & Swanson, C. E. (1998). Factors influencing wound healing after surgery for metastatic disease of the spine. *Spine (Phila Pa 1976), 23*, 726. https://doi.org/10.1097/00007632-199803150-00015.
30. Schuster, J. M., Rechtine, G., Norvell, D. C., & Dettori, J. R. (2010). The influence of perioperative risk factors and therapeutic interventions on infection rates after spine surgery: A systematic review. *Spine (Phila Pa 1976), 35*, S125. https://doi.org/10.1097/BRS.0b013e3181d8342c.
31. O'Toole, J. E., Eichholz, K. M., & Fessler, R. G. (2009). Surgical site infection rates after minimally invasive spinal surgery. *Journal of Neurosurgery. Spine, 11*, 471. https://doi.org/10.3171/2009.5.SPINE08633.
32. Parker, S. L., Adogwa, O., Witham, T. F., Aaronson, O. S., Cheng, J., & McGirt, M. J. (2011). Post-operative infection after minimally invasive versus open transforaminal lumbar interbody fusion (TLIF): Literature review and cost analysis. *Minimally Invasive Neurosurgery, 54*, 33. https://doi.org/10.1055/s-0030-1269904.
33. McGirt, M. J., Parker, S. L., Lerner, J., Engelhart, L., Knight, T., & Wang, M. Y. (2011). Comparative analysis of perioperative surgical site infection after minimally invasive versus open posterior/transforaminal lumbar interbody fusion: Analysis of hospital billing and discharge data from 5170 patients. *Journal of Neurosurgery. Spine, 14*, 771. https://doi.org/10.3171/2011.1.SPINE10571.
34. Ali, R., Schwalb, J. M., Nerenz, D. R., Antoine, H. J., & Rubinfeld, I. (2016). Use of the modified frailty index to predict 30-day morbidity and mortality from spine surgery. *Journal of Neurosurgery. Spine, 25*, 537. https://doi.org/10.3171/2015.10.SPINE14582.
35. Pull ter Gunne, A. F., Hosman, A. J. F., Cohen, D. B., Schuetz, M., Habil, D., van Laarhoven, C. J. H. M., et al. (2012). A methodological systematic review on surgical site infections following spinal surgery: part 1: risk factors. *Spine (Phila Pa 1976), 37*, 2017. https://doi.org/10.1097/BRS.0b013e31825bfca8.
36. Mok, J. M., Pekmezci, M., Piper, S. L., Boyd, E., Berven, S. H., Burch, S., et al. (2008). Use of C-reactive protein after spinal surgery: Comparison with erythrocyte sedimentation rate as predictor of early postoperative infectious complications. *Spine (Phila Pa 1976), 33*, 415. https://doi.org/10.1097/BRS.0b013e318163f9ee.
37. Choi, M. K., Kim, S. B., Kim, K. D., & Ament, J. D. (2014). Sequential changes of plasma c-reactive protein, Erythrocyte sedimentation rate and white blood cell count in spine surgery : Comparison between lumbar open discectomy and posterior lumbar interbody fusion. *Journal of Korean Neurosurgical Association, 56*, 218. https://doi.org/10.3340/jkns.2014.56.3.218.
38. Thelander, U., & Larsson, S. (1992). Quantitation of C-reactive protein levels and erythrocyte sedimentation rate after spinal surgery. *Spine (Phila Pa 1976), 17*, 400.
39. Deguchi, M., Shinjo, R., Yoshioka, Y., & Seki, H. (2010). The usefulness of serum amyloid A as a postoperative inflammatory marker after posterior lumbar interbody fusion. *Journal of Bone and Joint Surgery. British Volume (London), 92*, 555. https://doi.org/10.1302/0301-620X.92B4.22807.
40. Nie, H., Jiang, D., Ou, Y., Quan, Z., Hao, J., Bai, C., et al. (2011). Procalcitonin as an early predictor of postoperative infectious complications in patients with acute traumatic spinal cord injury. *Spinal Cord, 49*, 715. https://doi.org/10.1038/sc.2010.190.
41. Berbari, E., Mabry, T., Tsaras, G., Spangehl, M., Erwin, P. J., Murad, M. H., et al. (2010). Inflammatory blood laboratory levels as markers of prosthetic joint infection: A systematic review and meta-analysis. *The Journal of Bone and Joint Surgery. American Volume, 92*, 2102. https://doi.org/10.2106/JBJS.I.01199.
42. Sponseller, P. D., LaPorte, D. M., Hungerford, M. W., Eck, K., Bridwell, K. H., & Lenke, L. G. (2000). Deep wound infections after neuromuscular scoliosis surgery: A multicenter study of risk factors and treatment outcomes. *Spine (Phila Pa 1976), 25*, 2461. https://doi.org/10.1097/00007632-200010010-00007.

43. Tarabichi, M., Shohat, N., Goswami, K., Alvand, A., Silibovsky, R., Belden, K., et al. (2018). Diagnosis of periprosthetic joint infection: The potential of next-generation sequencing. *The Journal of Bone and Joint Surgery. American Volume, 100*, 147. https://doi.org/10.2106/JBJS.17.00434.
44. Bémer, P., Plouzeau, C., Tande, D., Léger, J., Giraudeau, B., Valentin, A. S., et al. (2014). Evaluation of 16S rRNA gene PCR sensitivity and specificity for diagnosis of prosthetic joint infection: A prospective multicenter cross-sectional study. *Journal of Clinical Microbiology, 52*, 3583. https://doi.org/10.1128/JCM.01459-14.
45. Gomez, E., Cazanave, C., Cunningham, S. A., Greenwood-Quaintance, K. E., Steckelberg, J. M., Uhl, J. R., et al. (2012). Prosthetic joint infection diagnosis using broad-range PCR of biofilms dislodged from knee and hip arthroplasty surfaces using sonication. *Journal of Clinical Microbiology, 50*, 3501. https://doi.org/10.1128/JCM.00834-12.
46. Jacovides, C. L., Kreft, R., Adeli, B., Hozack, B., Ehrlich, G. D., & Parvizi, J. (2012). Successful identification of pathogens by polymerase chain reaction (PCR)-based electron spray ionization time-of-flight mass spectrometry (ESI-TOF-MS) in culture-negative periprosthetic joint infection. *The Journal of Bone and Joint Surgery. American Volume, 94*, 2247. https://doi.org/10.2106/JBJS.L.00210.
47. Parvizi, J., & Gehrke, T. (2014). Definition of periprosthetic joint infection. *The Journal of Arthroplasty, 29*, 1331. https://doi.org/10.1016/j.arth.2014.03.009.
48. Parvizi, J., Tan, T. L., Goswami, K., Higuera, C., Della Valle, C., Chen, A. F., et al. (2018). The 2018 definition of periprosthetic hip and knee infection: An evidence-based and validated criteria. *The Journal of Arthroplasty, 33*, 1309. https://doi.org/10.1016/j.arth.2018.02.078.
49. Parvizi, J., Zmistowski, B., Berbari, E. F., Bauer, T. W., Springer, B. D., Della Valle, C. J., et al. (2011). New definition for periprosthetic joint infection: From the workgroup of the musculoskeletal infection society. *Clinical Orthopaedics and Related Research, 469*, 2992–2994. https://doi.org/10.1007/s11999-011-2102-9.
50. Silber, J. S., Anderson, D. G., Vaccaro, A. R., Anderson, P. A., & McCormick, P. (2002). Management of postprocedural discitis. *The Spine Journal, 2*, 279. https://doi.org/10.1016/S1529-9430(02)00203-6.
51. Inanami, H., Oshima, Y., Iwahori, T., Takano, Y., Koga, H., & Iwai, H. (2015). Role of 18F-Fluoro-D-deoxyglucose PET/CT in diagnosing surgical site infection after spine surgery with instrumentation. *Spine (Phila Pa 1976), 40*, 109–113. https://doi.org/10.1097/BRS.0000000000000674.
52. Dauchy, F.-A., Dutertre, A., Lawson-Ayayi, S., de Clermont-Gallerande, H., Fournier, C., Zanotti-Fregonara, P., et al. (2016). Interest of [18F]fluorodeoxyglucose positron emission tomography/computed tomography for the diagnosis of relapse in patients with spinal infection: a prospective study. *Clinical Microbiology and Infection, 22*, 438–443. https://doi.org/10.1016/J.CMI.2015.12.028.
53. De Winter, F., Gemmel, F., Van De Wiele, C., Poffijn, B., Uyttendaele, D., & Dierckx, R. (2003). 18-Fluorine Fluorodeoxyglucose Positron Emission Tomography for the Diagnosis of Infection in the Postoperative Spine. *Spine (Phila Pa 1976), 28*, 1314–1319. https://doi.org/10.1097/01.BRS.0000065483.07790.34.
54. Kim, S.-J., Lee, S. H., Chung, H. W., Lee, M. H., Shin, M. J., & Park, S. W. (2017). Magnetic resonance imaging patterns of post-operative spinal infection: relationship between the clinical onset of infection and the infection site. *Journal of Korean Neurosurgical Association, 60*, 448–455. https://doi.org/10.3340/jkns.2015.0505.010.
55. Djukic, S., Genant, H. K., Helms, C. A., & Holt, R. G. (1990). Magnetic resonance imaging of the postoperative lumbar spine. *Radiologic Clinics of North America, 28*, 341–360.
56. Boden, S. D., Davis, D. O., Dina, T. S., Sunner, J. L., & Wiesel, S. W. (1992). Postoperative diskitis: distinguishing early MR imaging findings from normal postoperative disk space changes. *Radiology, 184*, 765. https://doi.org/10.1148/radiology.184.3.1509065.
57. Kimura, H., Shikata, J., Odate, S., & Soeda, T. (2017). Pedicle screw fluid sign: An indication on magnetic resonance imaging of a deep infection after posterior spinal instrumentation. *Clinical Spine Surgery, 30*, 169. https://doi.org/10.1097/BSD.0000000000000040.

58. Diehn, F. E. (2012). Imaging of spine infection. *Radiologic Clinics of North America, 50*, 777–798. https://doi.org/10.1016/j.rcl.2012.04.001.
59. Ganesh, D., Gottlieb, J., Chan, S., Martinez, O., & Eismont, F. (2015). Fungal infections of the spine. *Spine (Phila Pa 1976), 40*, E719. https://doi.org/10.1097/BRS.0000000000000903.
60. Koutsoumbelis, S., Hughes, A. P., Girardi, F. P., Cammisa, F. P., Finerty, E. A., Nguyen, J. T., et al. (2011). Risk factors for postoperative infection following posterior lumbar instrumented arthrodesis. *The Journal of Bone and Joint Surgery. American, 93*, 1627. https://doi.org/10.2106/JBJS.J.00039.
61. Adogwa, O., Elsamadicy, A. A., Sergesketter, A., Vuong, V. D., Mehta, A. I., Vasquez, R. A., et al. (2017). Prophylactic use of intraoperative vancomycin powder and postoperative infection: An analysis of microbiological patterns in 1200 consecutive surgical cases. *Journal of Neurosurgery. Spine, 27*, 328. https://doi.org/10.3171/2017.2.SPINE161310.
62. Grabel, Z. J., Boden, A., Segal, D. N., Boden, S., Milby, A. H., & Heller, J. G. (2018). The impact of prophylactic intraoperative vancomycin powder on microbial profile, antibiotic regimen, length of stay, and reoperation rate in elective spine surgery. *The Spine Journal, 19*, 261. https://doi.org/10.1016/j.spinee.2018.05.036.
63. Butler-Wu, S. M., Burns, E. M., Pottinger, P. S., Magaret, A. S., Rakeman, J. L., Matsen, F. A., et al. (2011). Optimization of periprosthetic culture for diagnosis of Propionibacterium acnes prosthetic joint infection. *Journal of Clinical Microbiology, 49*, 2490. https://doi.org/10.1128/JCM.00450-11.
64. Weinstein, M. A., McCabe, J. P., & Cammisa, J. (2000). Postoperative spinal wound infection: A review of 2,391 consecutive index procedures. *Journal of Spinal Disorders, 13*, 422. https://doi.org/10.1097/00002517-200010000-00009.
65. Hahn, F., Zbinden, R., & Min, K. (2005). Late implant infections caused by Propionibacterium acnes in scoliosis surgery. *European Spine Journal, 14*, 783. https://doi.org/10.1007/s00586-004-0854-6.
66. Wimmer, C., Gluch, H., Franzreb, M., & Ogon, M. (1998). Predisposing factors for infection in spine surgery: a survey of 850 spinal procedures. *Journal of Spinal Disorders, 11*, 124–128.
67. Pulido, L., Ghanem, E., Joshi, A., Purtill, J. J., & Parvizi, J. (2008). Periprosthetic joint infection: The incidence, timing, and predisposing factors. *Clinical Orthopaedics and Related Research, 466*, 1710–1715. https://doi.org/10.1007/s11999-008-0209-4.
68. Durack, D. T. (1975). Experimental bacterial endocarditis. IV. Structure and evolution of very early lesions. *The Journal of Pathology, 115*, 81–89. https://doi.org/10.1002/path.1711150204.
69. Durack, D. T., & Beeson, P. B. (1972). Experimental bacterial endocarditis. II. Survival of a bacteria in endocardial vegetations. *British Journal of Experimental Pathology, 53*, 50–53.
70. Davies, D. G., & Geesey, G. G. (1995). Regulation of the alginate biosynthesis gene algC in Pseudomonas aeruginosa during biofilm development in continuous culture. *Applied and Environmental Microbiology, 61*, 860–867. doi:7793920.
71. Donlan, R. M., & Costerton, J. W. (2002). Biofilms: Survival mechanisms of clinically relevant microorganisms. *Clinical Microbiology Reviews, 15*, 167–193. https://doi.org/10.1128/CMR.15.2.167-193.2002.
72. Bernthal, N. M., Stavrakis, A. I., Billi, F., Cho, J. S., Kremen, T. J., Simon, S. I., et al. (2010). A mouse model of post-arthroplasty Staphylococcus aureus joint infection to evaluate in vivo the efficacy of antimicrobial implant coatings. *PLoS One, 5*, 1–11. https://doi.org/10.1371/journal.pone.0012580.
73. Thakkar, V., Ghobrial, G. M., Maulucci, C. M., Singhal, S., Prasad, S. K., Harrop, J. S., et al. (2014). Nasal MRSA colonization: Impact on surgical site infection following spine surgery. *Clinical Neurology and Neurosurgery, 125*, 94. https://doi.org/10.1016/j.clineuro.2014.07.018.
74. Kim, D. H., Spencer, M., Davidson, S. M., Li, L., Shaw, J. D., Gulczynski, D., et al. (2010). Institutional Prescreening for detection and eradication of methicillin-resistant

Staphylococcus aureus in patients undergoing elective orthopaedic surgery. *The Journal of Bone and Joint Surgery. American Volume, 92*, 1820.
75. Gruskay, J., Kepler, C., Smith, J., Radcliff, K., & Vaccaro, A. (2012). Is surgical case order associated with increased infection rate after spine surgery? *Spine (Phila Pa 1976), 37*, 1170. https://doi.org/10.1097/BRS.0b013e3182407859.
76. Gruskay, J., Smith, J., Kepler, C. K., Radcliff, K., Harrop, J., Albert, T., et al. (2013). The seasonality of postoperative infection in spine surgery. *Journal of Neurosurgery. Spine, 18*, 57. https://doi.org/10.3171/2012.10.SPINE12572.
77. Bible, J. E., O'Neill, K. R., Crosby, C. G., Schoenecker, J. G., McGirt, M. J., & Devin, C. J. (2013). Implant contamination during spine surgery. *The Spine Journal, 13*, 637. https://doi.org/10.1016/j.spinee.2012.11.053.
78. Mehta, A. I., Babu, R., Karikari, I. O., Grunch, B., Agarwal, V. J., Owens, T. R., et al. (2012). 2012 young investigator award winner: The distribution of body mass as a significant risk factor for lumbar spinal fusion postoperative infections. *Spine (Phila Pa 1976), 37*, 1652. https://doi.org/10.1097/BRS.0b013e318241b186.
79. Jackson, K. L., & Devine, J. G. (2016). The effects of obesity on spine surgery: A systematic review of the literature. *Global Spine Journal, 6*, 394. https://doi.org/10.1055/s-0035-1570750.
80. Jain, D., Berven, S. H., Carter, J., Zhang, A. L., & Deviren, V. (2018). Bariatric surgery before elective posterior lumbar fusion is associated with reduced medical complications and infection. *The Spine Journal, 18*, 1526. https://doi.org/10.1016/j.spinee.2018.01.023.
81. Golinvaux, N. S., Varthi, A. G., Bohl, D. D., Basques, B. A., & Grauer, J. N. (2014). Complication rates following elective lumbar fusion in patients with diabetes. *Spine (Phila Pa 1976), 39*, 1809–1816. https://doi.org/10.1097/BRS.0000000000000506.
82. Hikata, T., Iwanami, A., Hosogane, N., Watanabe, K., Ishii, K., Nakamura, M., et al. (2014). High preoperative hemoglobin A1c is a risk factor for surgical site infection after posterior thoracic and lumbar spinal instrumentation surgery. *Journal of Orthopaedic Science, 19*, 223. https://doi.org/10.1007/s00776-013-0518-7.
83. Yang, S., Werner, B. C., Cancienne, J. M., Hassanzadeh, H., Shimer, A. L., Shen, F. H., et al. (2016). Preoperative epidural injections are associated with increased risk of infection after single-level lumbar decompression. *The Spine Journal, 16*, 191. https://doi.org/10.1016/j.spinee.2015.07.439.
84. Cancienne, J. M., Werner, B. C., Puvanesarajah, V., Hassanzadeh, H., Singla, A., Shen, F. H., et al. (2017). Does the timing of preoperative epidural steroid injection affect infection risk after ACDF or posterior cervical fusion? *Spine (Phila Pa 1976), 42*, 71–77. https://doi.org/10.1097/BRS.0000000000001661.
85. Sidhwa, F., & Itani, K. M. F. (2015). Skin preparation before surgery: Options and evidence. *Surgical Infections, 16*, 14. https://doi.org/10.1089/sur.2015.010.
86. Savage, J. W., Weatherford, B. M., Sugrue, P. A., Nolden, M. T., Liu, J. C., Song, J. K., et al. (2012). Efficacy of surgical preparation solutions in lumbar spine surgery. *The Journal of Bone and Joint Surgery. American Volume, 94*, 490. https://doi.org/10.2106/JBJS.K.00471.
87. Berríos-Torres, S. I., Umscheid, C. A., Bratzler, D. W., Leas, B., Stone, E. C., Kelz, R. R., et al. (2017). Centers for Disease Control and Prevention guideline for the prevention of surgical site infection, 2017. *JAMA Surgery, 152*, 784. https://doi.org/10.1001/jamasurg.2017.0904.
88. Barker, F. G. (2002). Efficacy of prophylactic antibiotic therapy in spinal surgery: a meta-analysis. *Neurosurgery, 51*, 391. https://doi.org/10.1227/00006123-200208000-00017.
89. Bratzler, D. W., Dellinger, E. P., Olsen, K. M., Perl, T. M., Auwaerter, P. G., Bolon, M. K., et al. (2013). Clinical practice guidelines for antimicrobial prophylaxis in surgery. *American Journal of Health-System Pharmacy, 70*, 195. https://doi.org/10.2146/ajhp120568.
90. Anderson, D. J., Podgorny, K., Berríos-Torres, S. I., Bratzler, D. W., Dellinger, E. P., Greene, L., et al. (2014). Strategies to prevent surgical site infections in acute care hospitals: 2014 update. *Infection Control and Hospital Epidemiology, 35*, 605. https://doi.org/10.1086/676022.

91. Bull, A. L., Worth, L. J., & Richards, M. J. (2012). Impact of vancomycin surgical antibiotic prophylaxis on the development of methicillin-sensitive staphylococcus aureus surgical site infections: Report from Australian surveillance data (VICNISS). *Annals of Surgery, 256*, 1089. https://doi.org/10.1097/SLA.0b013e31825fa398.
92. Finkelstein, R., Rabino, G., Mashiah, T., Bar-El, Y., Adler, Z., Kertzman, V., et al. (2002). Vancomycin versus cefazolin prophylaxis for cardiac surgery in the setting of a high prevalence of methicillin-resistant staphylococcal infections. *The Journal of Thoracic and Cardiovascular Surgery, 123*, 326. https://doi.org/10.1067/mtc.2002.119698.
93. Reineke, S., Carrel, T. P., Eigenmann, V., Gahl, B., Fuehrer, U., Seidl, C., et al. (2018). Adding vancomycin to perioperative prophylaxis decreases deep sternal wound infections in high-risk cardiac surgery patients. *European Journal of Cardio-Thoracic Surgery, 53*, 428. https://doi.org/10.1093/ejcts/ezx328.
94. Tomov, M., Mitsunaga, L., Durbin-Johnson, B., Nallur, D., & Roberto, R. (2015). Reducing surgical site infection in spinal surgery with betadine irrigation and intrawound vancomycin powder. *Spine (Phila Pa 1976), 40*, 491–499. https://doi.org/10.1097/BRS.0000000000000789.
95. Brown, N. M., Cipriano, C. A., Moric, M., Sporer, S. M., & Della Valle, C. J. (2012). Dilute betadine lavage before closure for the prevention of acute postoperative deep periprosthetic joint infection. *The Journal of Arthroplasty, 27*, 27–30. https://doi.org/10.1016/j.arth.2011.03.034.
96. Rabenberg, V. S., Ingersoll, C. D., Sandrey, M. A., & Johnson, M. T. (2002). The bactericidal and cytotoxic effects of antimicrobial wound cleansers. *Journal of Athletic Training, 37*, 51–54.
97. Liu, J. X., Werner, J., Kirsch, T., Zuckerman, J. D., & Virk, M. S. (2018). Cytotoxicity evaluation of chlorhexidine gluconate on human fibroblasts, myoblasts, and osteoblasts. *Journal of Bone and Joint Infection, 3*, 165–172. https://doi.org/10.7150/jbji.26355.
98. Balin, A. K., & Pratt, L. (2002). Dilute povidone-iodine solutions inhibit human skin fibroblast growth. *Dermatologic Surgery, 28*, 210–214.
99. Sweet, F. A., Roh, M., & Sliva, C. (2011). Intrawound application of vancomycin for prophylaxis in instrumented thoracolumbar fusions: Efficacy, drug levels, and patient outcomes. *Spine (Phila Pa 1976), 36*, 2084. https://doi.org/10.1097/BRS.0b013e3181ff2cb1.
100. Strom, R. G., Pacione, D., Kalhorn, S. P., & Frempong-Boadu, A. K. (2013). Lumbar laminectomy and fusion with routine local application of vancomycin powder: Decreased infection rate in instrumented and non-instrumented cases. *Clinical Neurology and Neurosurgery, 115*, 1766. https://doi.org/10.1016/j.clineuro.2013.04.005.
101. O'Neill, K. R., Smith, J. G., Abtahi, A. M., Archer, K. R., Spengler, D. M., McGirt, M. J., et al. (2011). Reduced surgical site infections in patients undergoing posterior spinal stabilization of traumatic injuries using vancomycin powder. *The Spine Journal, 11*, 641. https://doi.org/10.1016/j.spinee.2011.04.025.
102. Zebala, L. P., Chuntarapas, T., Kelly, M. P., Talcott, M., Greco, S., & Riew, K. D. (2014). Intrawound vancomycin powder eradicates surgical wound contamination: An in vivo rabbit study. *The Journal of Bone and Joint Surgery. American Volume, 96*, 46. https://doi.org/10.2106/JBJS.L.01257.
103. Godil, S. S., Parker, S. L., O'Neill, K. R., Devin, C. J., & McGirt, M. J. (2013). Comparative effectiveness and cost-benefit analysis of local application of vancomycin powder in posterior spinal fusion for spine trauma: clinical article. *Journal of Neurosurgery. Spine, 19*, 331. https://doi.org/10.3171/2013.6.SPINE121105.
104. Gans, I., Dormans, J. P., Spiegel, D. A., Flynn, J. M., Sankar, W. N., Campbell, R. M., et al. (2013). Adjunctive vancomycin powder in pediatric spine surgery is safe. *Spine (Phila Pa 1976), 38*, 1703. https://doi.org/10.1097/BRS.0b013e31829e05d3.
105. Mendoza, M. C., Sonn, K. A., Kannan, A. S., Bellary, S. S., Mitchell, S. M., Singh, G., et al. (2016). The effect of vancomycin powder on bone healing in a rat spinal rhBMP-2 model. *Journal of Neurosurgery. Spine, 25*, 147–153. https://doi.org/10.3171/2015.11.SPINE15536.

106. Laratta, J. L., Shillingford, J. N., Hardy, N., Lombardi, J. M., Saifi, C., Romanov, A., et al. (2017). Intrawound tobramycin powder eradicates surgical wound contamination. *Spine (Phila Pa 1976), 42*, E1393. https://doi.org/10.1097/BRS.0000000000002187.
107. Han, J.-S., Kim, S.-H., Jin, S.-W., Lee, S.-H., Kim, B.-J., Kim, S.-D., et al. (2016). The use of gentamicin-impregnated collagen sponge for reducing surgical site infection after spine surgery. *Korean J Spine, 13*, 129–133. https://doi.org/10.14245/kjs.2016.13.3.129.
108. Scott, C. P., Higham, P. A., & Dumbleton, J. H. (1999). Effectiveness of bone cement containing tobramycin. An in vitro susceptibility study of 99 organisms found in infected joint arthroplasty. *Journal of Bone and Joint Surgery. British Volume (London), 81*, 440–443.
109. Scott, C. P., & Higham, P. A. (2003). Antibiotic bone cement for the treatment of pseudomonas aeruginosa in joint arthroplasty: Comparison of tobramycin and gentamicin-loaded cements. *Journal of Biomedical Materials Research, 64B*, 94–98. https://doi.org/10.1002/jbm.b.10515.
110. Emohare, O., Ledonio, C. G., Hill, B. W., Davis, R. A., Polly, D. W., & Kang, M. M. (2014). Cost savings analysis of intrawound vancomycin powder in posterior spinal surgery. *The Spine Journal, 14*, 2710. https://doi.org/10.1016/j.spinee.2014.03.011.
111. Youssef, J. A., Orndorff, D. G., Scott, M. A., Ebner, R. E., & Knewitz, A. P. (2014). Sterile Seroma resulting from multilevel XLIF procedure as possible adverse effect of prophylactic vancomycin powder: A case report. *Evidence Based Spine Care Journal, 5*, 127–133. https://doi.org/10.1055/s-0034-1386754.
112. Mariappan, R., Manninen, P., Massicotte, E. M., & Bhatia, A. (2013). Circulatory collapse after topical application of vancomycin powder during spine surgery. *Journal of Neurosurgery. Spine, 19*, 381–383. https://doi.org/10.3171/2013.6.SPINE1311.
113. Armaghani, S. J., Menge, T. J., Lovejoy, S. A., Mencio, G. A., & Martus, J. E. (2014). Safety of topical vancomycin for pediatric spinal deformity. *Spine (Phila Pa 1976), 39*, 1683–1687. https://doi.org/10.1097/BRS.0000000000000465.
114. Webster, J., & Alghamdi, A. (2015). Use of plastic adhesive drapes during surgery for preventing surgical site infection. *Cochrane Database Syst Rev*. https://doi.org/10.1002/14651858.CD006353.pub4.
115. Brown, M. D., & Brookfield, K. F. W. (2004). A randomized study of closed wound suction drainage for extensive lumbar spine surgery. *Spine (Phila Pa 1976), 29*, 1066. https://doi.org/10.1097/00007632-200405150-00003.
116. Diab, M., Smucny, M., Dormans, J. P., Erickson, M. A., Ibrahim, K., Lenke, L. G., et al. (2012). Use and outcomes of wound drain in spinal fusion for adolescent idiopathic scoliosis. *Spine (Phila Pa 1976), 37*, 966. https://doi.org/10.1097/BRS.0b013e31823bbf0b.
117. Kato, S., Chikuda, H., Ohya, J., Oichi, T., Matsui, H., Fushimi, K., et al. (2016). Risk of infectious complications associated with blood transfusion in elective spinal surgery—a propensity score matched analysis. *The Spine Journal, 16*, 55–60. https://doi.org/10.1016/j.spinee.2015.10.014.
118. Fisahn, C., Jeyamohan, S., Norvell, D. C., Tubbs, R. S., Moisi, M., Chapman, J. R., et al. (2017). Association between allogeneic blood transfusion and postoperative infection in major spine surgery. *Clin Spine Surg, 30*, E988–E992. https://doi.org/10.1097/BSD.0000000000000539.
119. Shiono, Y., Watanabe, K., Hosogane, N., Tsuji, T., Ishii, K., Nakamura, M., et al. (2012). Sterility of posterior elements of the spine in posterior correction surgery. *Spine (Phila Pa 1976), 37*, 523. https://doi.org/10.1097/BRS.0b013e318224d7b2.
120. Rehman, A., Rehman, A. U., Rehman, T. U., & Freeman, C. (2015). Removing outer gloves as a method to reduce spinal surgery infection. *Journal of Spinal Disorders & Techniques, 28*, E343. https://doi.org/10.1097/BSD.0b013e31829046ca.
121. Beldi, G., Bisch-Knaden, S., Banz, V., Mühlemann, K., & Candinas, D. (2009). Impact of intraoperative behavior on surgical site infections. *American Journal of Surgery, 198*, 157. https://doi.org/10.1016/j.amjsurg.2008.09.023.

122. Mangram, A. J., Horan, T. C., Pearson, M. L., Silver, L. C., & Jarvis, W. R. (1999). Guideline for prevention of surgical site infection, 1999. *American Journal of Infection Control, 27*, 97. https://doi.org/10.1016/S0196-6553(99)70088-X.
123. Arnold, W. V., Shirtliff, M. E., & Stoodley, P. (2014). Bacterial biofilms and periprosthetic infections. *Instructional Course Lectures, 63*, 385–391. https://doi.org/10.1016/S0021-9355(13)73979-1.
124. Lewis, K. (2010). Persister Cells. *Annual Review of Microbiology, 64*, 357–372. https://doi.org/10.1146/annurev.micro.112408.134306.
125. Endara, M., & Attinger, C. (2012). Using color to guide debridement. *Advances in Skin & Wound Care, 25*, 549–555. https://doi.org/10.1097/01.ASW.0000425936.94874.9a.
126. Dorafshar, A. H., Gitman, M., Henry, G., Agarwal, S., & Gottlieb, L. J. (2010). Guided surgical debridement: Staining tissues with methylene blue. *Journal of Burn Care & Research, 31*, 791–794. https://doi.org/10.1097/BCR.0b013e3181eed1d6.
127. Dipaola, C. P., Saravanja, D. D., Boriani, L., Zhang, H., Boyd, M. C., Kwon, B. K., et al. (2012). Postoperative infection treatment score for the spine (PITSS): Construction and validation of a predictive model to define need for single versus multiple irrigation and debridement for spinal surgical site infection. *The Spine Journal, 12*, 218. https://doi.org/10.1016/j.spinee.2012.02.004.
128. Fehring, T. K., Odum, S. M., Berend, K. R., Jiranek, W. A., Parvizi, J., Bozic, K. J., et al. (2013). Failure of irrigation and débridement for early postoperative periprosthetic infection knee. *Clinical Orthopaedics and Related Research, 471*, 250–257. https://doi.org/10.1007/s11999-012-2373-9.
129. Sherrell, J. C., Fehring, T. K., Odum, S., Hansen, E., Zmistowski, B., Dennos, A., et al. (2011). The chitranjan ranawat award: Fate of two-stage reimplantation after failed irrigation and débridement for periprosthetic knee infection. *Clinical Orthopaedics and Related Research, 469*, 18–25. https://doi.org/10.1007/s11999-010-1434-1.
130. Shaw, J. D., Miller, S., Plourde, A., Shaw, D. L., Wustrack, R., & Hansen, E. N. (2017). Methylene blue-guided debridement as an intraoperative adjunct for the surgical treatment of periprosthetic joint infection. *The Journal of Arthroplasty, 32*, 3718. https://doi.org/10.1016/j.arth.2017.07.019.
131. Parry, J. A., Karau, M. J., Kakar, S., Hanssen, A. D., Patel, R., & Abdel, M. P. (2017). Disclosing agents for the intraoperative identification of biofilms on orthopedic implants. *The Journal of Arthroplasty, 32*, 2501. https://doi.org/10.1016/j.arth.2017.03.010.
132. Coulthwaite, L., & Verran, J. (2009). Evaluation of in vivo denture plaque assessment methods. *British Dental Journal, 207*:E12; discussion 282-283., E12. https://doi.org/10.1038/sj.bdj.2009.854.
133. Carter, K., Landini, G., Damien Walmsley, A., & Walmsley, A. D. (2004). Automated quantification of dental plaque accumulation using digital imaging. *Journal of Dentistry, 32*, 623–628. https://doi.org/10.1016/j.jdent.2004.06.006.
134. Hedequist, D., Haugen, A., Hresko, T., & Emans, J. (2009). Failure of attempted implant retention in spinal deformity delayed surgical site infections. *Spine (Phila Pa 1976), 34*, 60. https://doi.org/10.1097/BRS.0b013e31818ed75e.
135. Núñez-Pereira, S., Pellisé, F., Rodríguez-Pardo, D., Pigrau, C., Bagó, J., Villanueva, C., et al. (2013). Implant survival after deep infection of an instrumented spinal fusion. *The Bone and Joint Journal, 95-B*, 1121. https://doi.org/10.1302/0301-620X.95B8.
136. Rihn, J. A., Lee, J. Y., & Ward, W. T. (2008). Infection after the surgical treatment of adolescent idiopathic scoliosis: Evaluation of the diagnosis, treatment, and impact on clinical outcomes. *Spine (Phila Pa 1976), 33*, 289. https://doi.org/10.1097/BRS.0b013e318162016e.
137. Kasliwal, M. K., L a, T., & Traynelis, V. C. (2013). Infection with spinal instrumentation: Review of pathogenesis, diagnosis, prevention, and management. *Surgical Neurology International, 4*, 392. https://doi.org/10.4103/2152-7806.120783.
138. Alpert, H. W., Farley, F. A., Caird, M. S., Hensinger, R. N., Li, Y., & Vanderhave, K. L. (2014). Outcomes following removal of instrumentation after posterior spinal fusion. *Journal of Pediatric Orthopedics, 34*, 1. https://doi.org/10.1097/BPO.0000000000000145.

139. Kanayama, M., Hashimoto, T., Shigenobu, K., Oha, F., Iwata, A., & Tanaka, M. (2017). MRI-based decision making of implant removal in deep wound infection after instrumented lumbar fusion. *Clinical Spine Surgery, 30*, E99. https://doi.org/10.1097/BSD.0b013e3182aa4c72.
140. McPherson, E. J., Dipane, M. V., & Sherif, S. M. (2013). Dissolvable antibiotic beads in treatment of Periprosthetic joint infection and revision arthroplasty - the use of synthetic pure calcium Sulfate (Stimulan®) impregnated with vancomycin & tobramycin. *Reconstructive Review, 3*, 32–43. https://doi.org/10.15438/rr.v3i1.27.
141. Hegde, V. (2012). Management of postoperative spinal infections. *World Journal of Orthopedics, 3*, 182. https://doi.org/10.5312/wjo.v3.i11.182.
142. Seligson, D., Mehta, S., Voos, K., Henry, S. L., & Johnson, J. R. (1992). The use of antibiotic-impregnated polymethylmethacrylate beads to prevent the evolution of localized infection. *Journal of Orthopaedic Trauma, 6*, 401–406.
143. Masuda, S., Fujibayashi, S., Otsuki, B., Kimura, H., & Matsuda, S. (2017). Efficacy of target drug delivery and dead space reduction using antibiotic-loaded bone cement for the treatment of complex spinal infection. *Clin Spine Surg, 30*, E1246. https://doi.org/10.1097/BSD.0000000000000567.
144. Zhu, E. S., Thompson, G. R., Kreulen, C., & Giza, E. (2013). Amphotericin B-impregnated bone cement to treat refractory coccidioidal osteomyelitis. *Antimicrobial Agents and Chemotherapy, 57*, 6341. https://doi.org/10.1128/AAC.00963-13.
145. Deelstra, J. J., Neut, D., & Jutte, P. C. (2013). Successful treatment of Candida Albicans-infected Total hip prosthesis with staged procedure using an antifungal-loaded cement spacer. *The Journal of Arthroplasty, 28*, 374.e5. https://doi.org/10.1016/j.arth.2012.04.034.
146. Walker, B., Koerner, J., Sankarayanaryanan, S., & Radcliff, K. (2014). A consensus statement regarding the utilization of BMP in spine surgery. *Current Reviews in Musculoskeletal Medicine, 7*, 208. https://doi.org/10.1007/s12178-014-9224-0.
147. O'Shaughnessy, B. A., Kuklo, T. R., & Ondra, S. L. (2008). Surgical treatment of vertebral osteomyelitis with recombinant human bone morphogenetic protein-2. *Spine (Phila Pa 1976), 33*, E132. https://doi.org/10.1097/BRS.0b013e3181657ee3.
148. Allen, R. T., Lee, Y. P., Stimson, E., & Garfin, S. R. (2007). Bone morphogenetic protein-2 (BMP-2) in the treatment of pyogenic vertebral osteomyelitis. *Spine (Phila Pa 1976), 32*, 2996. https://doi.org/10.1097/BRS.0b013e31815cde3e.
149. Ploumis, A., Mehbod, A. A., Dressel, T. D., Dykes, D. C., Transfeldt, E. E., & Lonstein, J. E. (2008). Therapy of spinal wound infections using vacuum-assisted wound closure: Risk factors leading to resistance to treatment. *Journal of Spinal Disorders & Techniques, 21*, 320. https://doi.org/10.1097/BSD.0b013e318141f99d.
150. Lee, R., Beder, D., Street, J., Boyd, M., Fisher, C., Dvorak, M., et al. (2018). The use of vacuum-assisted closure in spinal wound infections with or without exposed dura. *European Spine Journal, 27*, 2536. https://doi.org/10.1007/s00586-018-5612-2.
151. Yuan-Innes, M. J., Temple, C. L., & Lacey, M. S. (2001). Vacuum-assisted wound closure: a new approach to spinal wounds with exposed hardware. *Spine (Phila Pa 1976), 26*, E1.
152. Argenta, L. C., & Morykwas, M. J. (1997). Vacuum-assisted closure: a new method for wound control and treatment: Clinical experience. *Annals of Plastic Surgery, 38*, 563.
153. Dumanian, G. A., Ondra, S. L., Liu, J., Schafer, M. F., & Chao, J. D. (2003). Muscle flap salvage of spine wounds with soft tissue defects or infection. *Spine (Phila Pa 1976), 28*, 1203. https://doi.org/10.1097/01.BRS.0000067260.22943.48.
154. Hochberg, J., Ardenghy, M., Yuen, J., Gonzalez-Cruz, R., Miura, Y., Conrado, R. M., et al. (1998). Muscle and musculocutaneous flap coverage of exposed spinal fusion devices. *Plastic and Reconstructive Surgery, 102*, 385. https://doi.org/10.1097/00006534-199808000-00013.
155. Mericli, A. F., Tarola, N. A., Moore, J. H., Copit, S. E., Fox, J. W., & Tuma, G. A. (2010). Paraspinous muscle flap reconstruction of complex midline back wounds: Risk factors and postreconstruction complications. *Annals of Plastic Surgery, 65*, 219. https://doi.org/10.1097/SAP.0b013e3181c47ef4.

156. Mericli, A. F., Mirzabeigi, M. N., Moore, J. H., Fox, J. W., Copit, S. E., & Tuma, G. A. (2011). Reconstruction of complex posterior cervical spine wounds using the paraspinous muscle flap. *Plastic and Reconstructive Surgery, 128*, 148. https://doi.org/10.1097/PRS.0b013e3182174075.
157. Marschall, J., Lane, M. A., Beekmann, S. E., Polgreen, P. M., & Babcock, H. M. (2013). Current management of prosthetic joint infections in adults: Results of an emerging infections network survey. *International Journal of Antimicrobial Agents, 41*, 272. https://doi.org/10.1016/j.ijantimicag.2012.10.023.
158. Keller, S. C., Cosgrove, S. E., Higgins, Y., Piggott, D. A., Osgood, G., & Auwaerter, P. G. (2016). Role of suppressive oral antibiotics in orthopedic hardware infections for those not undergoing two-stage replacement surgery. *Open Forum Infectious Diseases, 3*, ofw176. https://doi.org/10.1093/ofid/ofw176.
159. Kim, B. N., Kim, E. S., & Oh, M. D. (2014). Oral antibiotic treatment of staphylococcal bone and joint infections in adults. *The Journal of Antimicrobial Chemotherapy, 69*, 309–322. https://doi.org/10.1093/jac/dkt374.

Targeting Biofilms in Translational Research

Nicholas N. Ashton and Dustin L. Williams

Abstract Biofilms underpin the disease etiology of nearly all opportunistic bacterial infections especially when integumentary barriers are surgically breached and foreign materials remain implanted. Endogenous spread from the patient's own microbial flora is the likely source of most surgical site infections. Biomaterials potentiate infection by providing a substrate for biofilm formation. The biofilm protects these pathogens from both host immunity and clinical interventions in a variety of ways. Biofilm-forming bacteria excrete sticky exopolysaccharides to form cohesive communal aggregates and adhesive attachments to foreign surfaces like devitalized tissues and implanted biomaterials; this strategy deranges phagocytic clearance by host immune cells. Quiescent phenotypic variants in the biofilm cells are tolerant of antibiotic concentrations many orders of magnitude greater than would otherwise kill planktonic phenotypes, concentrations greatly exceeding toxic thresholds bounding safe systemic antibiotic concentrations. Biofilms are, thus, a nidus for infection as tolerant cells can outlast clinical antibiotic courses to subsequently reseed infection. Much emphasis has been placed on preventing biofilm infections from occurring as clinical strategies for eradicating established biofilm infections frequently fail. Biofilm infections usually require extensive surgical intervention to remove implanted biomaterials and debride affected tissues. These procedures are costly and usually

N. N. Ashton (✉)
George E. Wahlen Department of Veterans Affairs, Salt Lake City, UT, USA

Department of Orthopaedics, University of Utah, Salt Lake City, UT, USA

D. L. Williams
George E. Wahlen Department of Veterans Affairs, Salt Lake City, UT, USA

Department of Orthopaedics, University of Utah, Salt Lake City, UT, USA

Department of Pathology, University of Utah, Salt Lake City, UT, USA

Department of Biomedical Engineering, University of Utah, Salt Lake City, UT, USA

Department of Physical Medicine and Rehabilitation, Uniformed Services University of the Health Sciences, Bethesda, MD, USA

D. L. Williams (ed.), *Targeting Biofilms in Translational Research, Device Development, and Industrial Sectors*,
https://doi.org/10.1007/978-3-030-30667-0_9

accompanied by high patient mortality and morbidity. There is a pressing need for strategies that specifically target the biofilm because clinical measures to prevent infection still fail in this antibiotic era at great financial and physical expense.

Keywords Targeting biofilms · Translational research · Initial inocula · Antibiotic · Tolerance · Infection

Sources of Infection

In 1975 the clinicians Krizek and Robson poignantly suggested that "infection is less likely some process of contagion or bacterial visitation from without than it is a disturbance in the delicate balance man has established with his bacterial environment." Our growing understanding of surgical site infection, now in 2019, still supports this assertion. Opportunist pathogenic bacteria might arrive at the surgical site from various exogenous and endogenous origins, including healthcare personnel, airborne particles, and surgical instruments, from preexisting infection, or from the patient's own microbiome; yet now, subsequent to the advances of antiseptic and aseptic techniques formalized in the nineteenth century, endogenous culprits are suspected as the primary source of most surgical site infections where the protective skin barrier has not been traumatically compromised [1, 2]. As evidence, the most common infectious agents across nearly all procedure types are conspicuously represented in the local flora of the operation site. Most surgical procedures require percutaneous access through a dermal incision; in these, opportunistic organisms from the skin microbiota are the typical culprits if infection does occur. The skin is predominantly colonized by various species of *Staphylococcus*, *Corynebacterium*, *Streptococcus*, *Pseudomonas*, and *Cutibacterium* [3]. Many of these organisms can become pathogenic and are, therefore, categorized as opportunistic pathogens [3].

An analysis of 44 infected total knee arthroplasties presents one illustrative example of otherwise commensal skin bacteria becoming pathogenic; staphylococcal species like *S. aureus* and *S. epidermidis* were the most abundant causative agents at a combined ~43% with lesser contributions from streptococcus and gram-negative bacilli like *Pseudomonas aeruginosa* and *Proteus mirabilis* [4]. The skin flora is broadly represented in many other orthopedic infection studies in both total joint arthroplasties and fracture fixation procedures: *S. aureus* (20–30%); coagulase negative staphylococcal species, usually *S. epidermidis* (20–40%); streptococci (1–10%); gram-negative bacilli, usually *Pseudomonas* and sometimes enterobacteriaceae (6–17%); and small amounts of *Cutibacterium* (<5%) [5]. Percutaneous vascular procedures including coronary intervention and placement of vascular catheters are similarly prone to infection with bacteria of the skin flora. The most common reported pathogens in these vascular procedures were *S. aureus* (40–80%) followed by *P. aeruginosa* (16–20%) [6, 7].

Skin flora varies by anatomical location; these variations alter the pathogen profile of associated infections. *C. acnes* (previously *Propionibacterium acnes*), the most common causative agent of acne vulgaris, occupies the pilosebaceous follicles, which are in highest density in skin of the scalp, shoulders, back, upper chest, and face [8, 9]. Indeed, percutaneous procedures performed in these anatomical regions display much higher ratios of infection with *C. acnes* than at other percutaneous sites [10]. Follow-up analysis of 1571 primary shoulder prostheses revealed *C. acnes* as the most common infectious agent at 38%, surpassing *S. aureus* [11]. These results mirrored an earlier study finding, "The proportion of patients with shoulder infection who had infection due to *P. acnes* was significantly greater than the proportion of patients with lower limb infection who had infection due to *P. acnes* (9 of 16 patients vs. 1 of 233 patients; P<.001)." [12] An abundance of *C. acnes* infections was also observed in breast augmentation procedures where the authors noted, "*Propionibacterium* species were the microorganisms most frequently isolated from breast implants." [13] Likewise, in a study of 112 retrieved spinal implants, *P. acnes* was the most frequently detected microorganism (45%) followed by coagulase-negative staphylococci (40%) [14].

The unique bacterial flora of nondermal sites are also well-represented in infection profiles ranging from gynecological to gastrointestinal surgical procedures, thus further implicating endogenous culprits in most surgical site infections [15]. In a multicenter cohort study including 610 patients with infection from gastrointestinal resection, 301 had "bowel-derived infections" defined by a causative agent typical to the bowl flora: gram-negative bacilli, *Enterococcus* species, or anaerobic organisms [16]. Likewise, gynecological surgical sites had higher proportions of infection from the local flora including gram-negative bacilli, enterococci, and group B streptococci because of surgical incision across the vaginal wall and perineum [17].

Several authors have suggested that microorganisms can be found at ~90% of surgical sites upon wound closure; this upper-end estimate is reasonable [18, 19]. In a prospective of 66 patients undergoing open heart surgery, 47 (71%) had positive cultures from at least 1 swabbed location prior to closure [20]. Furthermore, sterile surgical instruments used in procedures in the body cavity were shown to have bioburden levels of ~10^2 per instrument, the contamination mostly attributed to microorganisms from the patient [21]. In one study, otherwise sterile pedicle screws were found in every procedure analyzed (n = 26) to have culturable bacteria after handling for placement by the surgeon [22]. Revised estimates for the number of human and bacterial cells in the body suggest that bacterial cells are as numerous as human cells and comprise a total mass of 0.2 kg for a reference person weighing 70 kg [23]. With current surgical approaches, there is a near inevitable spread of commensal organisms to the surgical site either within the timeframe of the procedure or from ensuing perioperative endogenous spread. Whether a successive infection does occur depends on many factors which can shift the balance toward either host defense mechanisms or the bacterial pathogen.

This apparent relationship between infection outcomes and the local flora of the surgery site suggests continued development of antiseptic techniques for eliminating surgery site microflora might produce more substantial gains in infection prevention compared with the development of aseptic procedures albeit both with diminishing returns (see chapter "We Begin to Target the Biofilm"). Beyond surgical scrubs, several products are on the market for preventing endogenous microflora from migrating to surgical sites and respective implants. Two examples, Biopatch® and GuardIVa® antimicrobial dressings, are impregnated with the broad-spectrum antimicrobial chlorhexidine gluconate. These devices are designed to dress and protect percutaneous catheter access sites. They have been shown effective at decreasing infection and reducing bacterial colonization of catheters [24, 25]. Devices like these, which target endogenous microflora culprits at the surgical site, will continue to improve infection outcomes.

Quantitative Bacteriology and the 10^5 Rule

Quantitative bacteriology has provided indispensable insight into the nature of surgical site infection and the mechanisms of its progression [26]. Possibly the first examples of clinically applied quantitative bacteriology were the use of cultures taken by French surgeons in WWI to find the bioburden in wounds older than 15 h; these cultures were then used to determine treatment course and closure time for the respective wounds [26–28]. By the mid-1900s, quantitative bacteriology was used experimentally to create and analyze wounds in animals; this enabled a more precise study of infection progression. Among other goals, these early researchers worked to find the minimum infectious doses of common pathogens across multiple tissue types. They used these newly developed animal models to study interventions at various stages of infection development. Minimum infectious doses vary significantly across bacterial species starting as low as 10^1 CFUs for the most pathogenic organisms like *Yersinia pestis*, *Giardia lamblia*, and *Shigella* [29–31]. At the other extreme are organisms considered both commensal and opportunistic pathogens as they are part of the natural human microbiota and also predominate in infected surgery sites and traumatic wounds. These display minimum infectious doses exceeding 10^5 CFUs including opportunistic organisms from *Staphylococcus*, *Streptococcus*, *Pseudomonas*, and *Cutibacterium*, among others [26, 32].

Most of this early animal work focused on the opportunistic pathogens typical of wound infections [32, 33]. Their techniques were ultimately applied to the skin of human volunteers, finding the minimum average pus-forming dose of ~7.5×10^6 cocci for staphylococcal organisms. This was one of the first in a series of quantitative observations in human tissues to associating a bioburden greater than 10^5 CFUs/ml or CFUs/g of tissue with infection. That same year in 1956, the analysis of urine from 74 patients "in whom the diagnosis of pyelonephritis (kidney infection) was made or suspected, 95 per cent were found to have more than 100,000 (10^5) bacteria per ml." [34, 35] In a study of 93 delayed primary wound closures, the authors observed a "quantitative relationship between bacterial contamination and clinical infection," noting a 96% successful closure rate if the bioburden was less than 10^5 CFU per g of tissue [36]. Results of these first wound-healing experiments mirrored those of an earlier rabbit study which showed skin grafts to fail if inoculated with streptococci, staphylococci, or *P. aeruginosa* at concentrations above 10.4×10^6 CFU/ml [37]. The significance of bioburdens in excess of 10^5 CFU/gram of tissue in infection outcomes has been subsequently observed in numerous animal and human studies; the clinical importance of the 10^5 CFUs/ml threshold of infection has been deemed "the 10^5 guideline" or "10^5 rule." [26, 38] Whether applicable or not, the 10^5 rule now pervades clinical laboratory protocols for testing antimicrobial compounds and, more generally, for testing antiinfection technologies, coatings, and devices.

The Exception to the 10^5 Rule

A prescient exception to the 10^5 rule was reported in 1957 by Elek and Conen. They showed that the implantation of one of the few widely used biomaterials of the day, a single silk suture, in the arm of a human volunteer would decrease the minimum pus-forming dose of *S. pyogenes* by 10^{-4} from 10^6 to 10^2 [39]. The authors commented, "The presence of a foreign body reaction in the form of sutures however resulted in a dramatic reduction of the minimum inoculum required to produce pus." [39] Aspects of this experiment were repeated in 1961 in the skin of mice where it was similarly observed that most mice would develop abscesses if an inoculum of 10^3 staphylococci was introduced on sutures; yet abscess formation occurred in a large portion of animals with lower inoculum of just 10–100 cocci [40]. The motivation for this study was, in-part, from common contemporary clinical observations that "Wound infection often begins about sutures with the formation of stitch abscesses." [40] The results of these early quantitative bacteriology studies would have been, in principle, unsurprising at the time as there was popular sentiment that foreign bodies worsen infection, and wound debridement was commonly employed as an antiinfection measure. As early as 1564, the French surgeon Ambroise Paré refuted the commonplace misconception of the time that gunshot wounds were poisons by showing the removal of the bullet, proper debridement, and antiseptic treatment could prevent infection. The remarkable aspect of the aforementioned quantitative bacteriology studies was the sheer extent to which the foreign suture material was shown to potentiate infection from even very low-level inocula.

These early observations of artificial materials as foci of infection serendipitously coincide with the development of numerous prostheses of artificial biomaterials which are now commonplace in clinical medicine [41]. For example, experimental vascular grafts sewn from silk handkerchiefs, parachute nylon, Terylene (polyester fabric), and Orlon (acrylic fabric) were first implanted in humans starting in 1952. The first two artificial heart valves with components of silicone, polycarbonate, and nylon were inserted in humans in 1953 and 1963. The first successful metal prosthetic hip joint was developed around 1958, preceding the first total knee prostheses by over a decade (1968–1972). In this period implanted biomaterials were required for other new medical technologies: intraocular lenses (1949), fully implantable pacemakers (1959), and the artificial kidney (1960), among others [41]. These events marked the start of a precipitous rise in clinically implanted biomaterials. These first observations of suture biomaterials potentiating infection in humans and animals (1957–1961) foreshadowed the most problematic and costly types of infection observed in the clinic today; contemporary clinical medicine now strongly relies on implanted biomaterials, more than ever, to save and improve the quality of human life. These difficult-to-treat infections are now commonly known as device-related infections and constitute the greatest burden to the healthcare system of all other infection types combined.

Biofilms Potentiate Infection

An understanding for how biomaterials potentiate infection was formalized by Costerton and colleagues from observations made in the 1970s [42]. They recognized that most bacteria, including opportunistic pathogens, preferentially dwell in sessile communities by excreting sticky extracellular polymeric substances (EPS) to form cohesive aggregates and adhesive attachments to implant biomaterials and devitalized tissues (e.g., sequestrum and necrotic soft tissue). These matrix-embedded communities are now commonly called bacterial biofilms. The hydrated EPS matrix encompassing the biofilm community comprises species-specific hydrophilic polymers, usually acidic polysaccharides, proteins, glyocproteins, glycolipids, and DNA. Synthesis of the EPS matrix is metabolically expensive yet provides a survival advantage across the varied bacterial niches of diverse organisms: from environmental bacteria in the soil and waterways to the common clinically relevant infectious opportunistic pathogens.

The biofilm strategy protects constituent bacteria from predation and attack by phagocytes; these are single-celled or multinucleated eukaryotes: protozoa, such as amoebas, in environmental systems or motile leucocytes, such as neutrophils, macrophages, or foreign-body giant cells in human tissues. Neutrophils are the most predominant polymorphonuclear leukocytes in the human tissues (~70%) and serve as the first line of defense against invading infectious bacteria [43, 44]. Although many pathogenic bacteria have developed a range of complex species-specific chemical signals and toxins to evade phagocytosis and clearance by host leucocytes [44, 45], the biofilm strategy broadly enables all biofilm-forming pathogens to evade phagocytic clearance through mechanical means. Biofilms can shroud constituent bacteria from opsonizing antibodies [46]. Leukocyte movement and chemotaxis might also be impaired in the viscous biofilm EPS matrix [47], an observation extrapolated from experiments showing impaired mobility of neutrophils in mucus and collagen gels [48–50]. Benchtop experiments showed that the EPS of mature *P. aeruginosa* biofilms reduced the chemotactic migration of neutrophils; the authors concluded that in the presences of biofilms, neutrophils "...lose their capacity to sense the direction and just slide over the EPS in a disoriented manner." [51] Biofilms mechanically frustrate engulfment by host phagocytes, which are unable to gain access to individual bacterium embedded in the EPS matrix [47, 52]. Neutrophils cannot engulf particles much larger than their own diameter (10–12 μm) as has been shown through experimentation with antibody-coated synthetic beads [53]. A representative 48 h in vitro staphylococcal biofilm is shown on implant titanium in Fig. 1. Biofilm growth studies with representative strains of pathogenic *S. aureus* and *S. epidermidis* have shown these organisms can rapidly colonize implant materials in only a few hours [54] and within just 4 h, nascent biofilm aggregates begin to frustrate neutrophil phagocytosis [54]. The biofilm, thus, affords indiscriminate protection from host clearance mechanisms even if the offending bacteria have not acquired other virulence factors.

Antibiotic Tolerance of Bacterial Biofilms

Antimicrobial compounds have antagonized microorganisms long before the industrial scale production and clinical use of antibiotics and antiseptics by humans. Most clinical antibiotics were either identified from natural sources (e.g., other microorganisms and fungi) or inspired from the structures of natural compounds [55]. Microorganisms have been locked in a perpetual cycle of ecological one-upmanship evolving a preponderance of sophisticated compounds to prevent the proliferation of competitors. The word antibiotic was contrived by Nobel Laureate Selman Waksman in 1941 to describe precisely this concept: "An antibiotic is a chemical substance, produced by micro-organisms, which has the capacity to inhibit the

Fig. 1 Scanning electron micrograph of a *S. aureus* ATCC 6538 biofilm grown in a CDC biofilm reactor for 48 h on medical grade titanium implant material

growth of and even to destroy bacteria and other micro-organisms." [55] Hijacking these sophisticated molecules to disrupt bacterial biosynthesis and metabolism of infectious pathogens now constitutes one of the greatest medical innovations of the twentieth century. Yet the biofilm mode of growth enables constituent bacteria to indiscriminately survive the highest clinical doses of antibiotic compounds even in strains that lack specific resistance genes. This survival mechanism is likely, in part, the result of selective evolutionary pressures, imparted microbe-against-microbe, through eons of biochemical warfare. The broad tolerance to antimicrobials observed in bacteria living in biofilm communities is primarily the consequence of phenotypic changes rather than the genotypic variation underpinning the antibiotic resistance epidemic; although in early biofilm literature, the terms antibiotic tolerance and antibiotic resistance were used interchangeably.

As an illustrative example, antibiotic kill profiles performed in our laboratory against methicillin-resistant *S. aureus* biofilms are shown in Fig. 2a and compared with the minimum inhibitory concentrations (MIC) against the planktonic phenotype in Fig. 2b. The biofilms exhibited strong recalcitrance to the clinical antibiotics gentamicin and vancomycin, which were unable to eradicate the biofilm at concentrations exceeding 500x the effective concentrations in planktonic phenotypes, although the upper margins of this experiment were not explored, and follow-on work is needed to compare equivalent starting concentrations of bacteria (MIC analyses begin with 10^5 CFU/ml; biofilm analyses often begin with 10^8 or more CFU/ml). The estimates in the literature for antibiotic tolerance in biofilms vary greatly ranging from 10–10,000× the effective concentrations against planktonic phenotypes. Unpublished antibiotic kill profiles against *S. aureus* ATCC 6538 were performed in our laboratory exploring the upper antibiotic tolerance limits of lab-grown biofilms. In these tests, vancomycin was ultimately pushed to its solubility limit in the test medium (64 mg/ml). At this inordinate concentration vancomycin produced less than 0.5 $\log_{10}$ reduction against a biofilm comprising >9 billion CFUs on both faces of a 0.5 in. coupon. Similarly, at 64 mg/ml gentamicin performed only slightly better with an approximate 3 $\log_{10}$ reduction (Fig. 2). Vancomycin and gentamicin were investigated as representative clinical standards because they are extensively used for staphylococcal infections of traumatic skin and orthopedic wounds and display clinically acceptable low MIC values against the *S. aureus* strains used. Both clinical antibiotics failed to eradicate *S. aureus* biofilms even at inordinate concentrations, over 60,000× the MIC, 6,000× the nephrotoxic and ototoxic thresholds, and 25–50× the cytotoxic limit.

The recalcitrance of biofilms to antibiotic and antimicrobial treatment was recognized early in the seminal work on biofilms [56, 57]. Costerton and colleagues suggested two possible mechanisms of biofilm tolerance to antibiotics: the first, "the limitation of [antibiotic] diffusion by the polyanionic matrix layer" and the second, "phenotypic adaptations to biofilm growth that alter the metabolic targets of these antibacterial agents." Overwhelming evidence now points to the latter hypothesis as the primary reason for antibiotic tolerance observed in biofilms. The growth of biofilm cells is constrained, in principle, by the same diffusion limitations of human cells, which are typically no more than 50 μm away from the nearest capillary, the

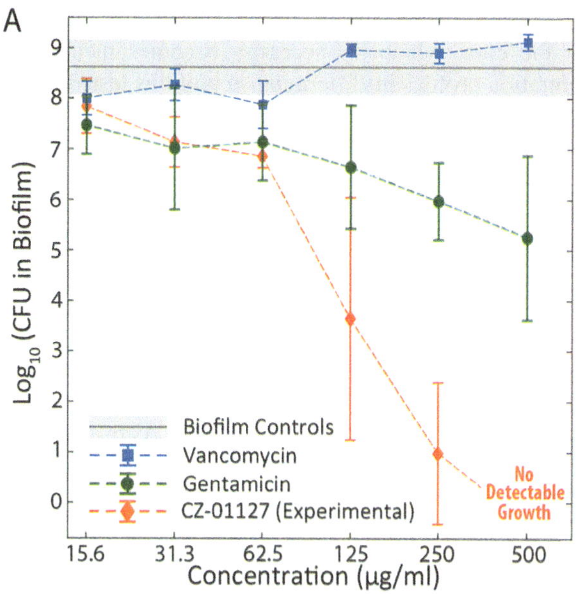

B

	MIC (μg/ml)	Cytotoxci Range (μg/ml)
Vancomycin	1	–
Gentamicin	1	>2000
CZ-01127	1	>125
CHG	2	>12.5

Fig. 2 (**a**) Antimicrobial efficacy of antibiotic solutions containing vancomycin (blue), gentamicin (green), and the experimental antibiotic CZ-01127 (red) against established biofilms of a clinical MRSA isolate grown in a CDC biofilm reactor. Baseline growth (black line) was determined for comparison. Error bars and gray-shaded region represent the standard deviation ($n = 5$ repeats each). (**b**) A table showing the MIC of CZ-01127 for comparison with those of vancomycin, gentamicin, and chlorhexidine gluconate (CHG) against the same MRSA isolate as in panel (**a**). The minimum passing values for a cytotoxic MEM elution assay are shown in the second column. These data are adapted with permission from Ashton et al. [65]

approximate width of a human hair. Biofilms do not have the sophisticated active circulatory systems of human tissues yet have been observed in some instances to develop reticulated channels and plumes enabling convective nutrient exchange. The biofilm relies on passive Fickian diffusion for delivery of oxygen and nutrients from the biofilm surface to the bacteria within [58]. In this simple system, diffusion time (t) is proportional to the square of diffusion distance (x) by Eq. 1, where D is the diffusion coefficient of oxygen or nutrients:

$$t \approx \frac{x^2}{2D} \quad (1)$$

From Eq. 1, as a biofilm thickens, oxygen and nutrients rapidly become diffusion limited within the core, which by necessity becomes metabolically quiescent. Oxygen concentration profiles in *P. aeruginosa* biofilms tested in air (20.95% oxygen) and 100% oxygen were shown experimentally to approach zero at depths of just 50 and 100 μm, respectively. Oxygen tensions in healthy human tissues are much lower from 1% to 11% and in the surgical site can dive even lower as compromised vasculature and infection will both intensify tissue hypoxia. Heterogeneous growth and phenotypic expression in biofilms is, in large part, due to the diffusion limitations of nutrients and oxygen into the interior of the biofilm. The outer metabolically active cells rapidly screen available nutrients leaving the internal cells in an anoxic metabolically quiescent state by necessity; it is these cells which primarily account for the antibiotic tolerance observed in biofilms. It is difficult to "gum up" the metaphorical cogs of bacterial cellular machinery with antibiotic compounds if the target cellular machinery is not turning. As discussed above, the most common infectious biofilm-forming pathogens implicated in device-related infections are conspicuously facultative anaerobes capable of both aerobic and anaerobic metabolism: *S. aureus*, *S. epidermidis*, *P. aeruginosa*, *C. acnes*, *E. coli*, and *Streptococcus* sp. Microorganisms are particularly bad at conserving genes which are not routinely utilized [59]. It is the authors' belief that the ubiquitous conservation of the cellular machinery required for anaerobic metabolism is owed to the biofilm mode of growth which predominates in natural ecosystems over planktonic modes.

There is an apparent relationship between the amount of anoxic nutrient-limited zones within a biofilm and the biofilm's tolerance to antibiotics. A thorough meta-analysis of literature data was performed to understand the factors associated with biofilm tolerance to biocides and antibiotics [60]. Numerous factors were investigated for their effect on antibiotic tolerance: molecular weight of the antimicrobial compound, substrate material on which the biofilm was grown, biofilm areal cell density, and biofilm age. If poor antibiotic penetration into the biofilm were the primary mechanism of antibiotic tolerance in biofilms, one would expect the molecular weight or diffusion coefficient of the antibiotic compound correlated with antibiotic tolerance; yet antibiotic tolerance in biofilms only correlated with areal cell density and age of the biofilm; in other words, larger and/or older mature biofilms can tolerate higher concentrations of antibiotics [61]. We observe evidence pointing in the same direction in our tests (like that shown in Fig. 2). A plateau is observed beyond a certain concentration for clinical antibiotics like vancomycin, gentamicin, and nafcillin, where additional increases, even up to the solubility limit of some antibiotics, have no additional consequence on viable bacteria numbers in the treated biofilm. This suggests a saturation effect where a population of cells express "phenotypic adaptations to biofilm growth that alter the metabolic targets of these antibacterial agents" as proposed by Costerton and colleagues years ago [56, 62].

Failures of Relying Exclusively on Planktonic Bacteria for Testing Antibiotics

Convention has it that on Friday September 28, 1928, Alexander Fleming began the discovery of the antibiotic penicillin by an astute observation after a blue-green sporulating mold contaminated one of his culture plates producing a zone of inhibition where the bacterial culture could not grow [63]. The exclusion criteria used in this groundbreaking observation was antagonized replication. After isolating the growth-inhibiting agent, penicillin, Fleming proceeded to refine a laboratory test to better quantify the degree to which this new compound might prevent replication of a given microorganism. He explained, "The inhibitory power can be accurately titrated by making serial dilutions of penicillin in fresh nutrient broth, and then implanting all the tubes with the same volume of bacterial suspension and incubating them. The inhibition can then readily be seen by noting the opacity of the broth." [64] This technique was further refined, in practice, giving rise to the minimum inhibitory concentration assay (MIC) which is unchanged in principle from Fleming's assay. Thus, the MIC is simply a measure of how effective a given compound is at preventing replication of a 10^5 CFU/ml planktonic culture. The 10^5 CFU/ml constraint was added because of the 10^5 CFU/ml guideline which is an approximate minimum infectious dose of bacteria for tissues that do not hold implanted biomaterials (see the preceding sections of this chapter) [38].

The minimum bactericidal concentration (MBC) is determined from the same dilution series as the MIC assay by transferring dilutions to agar plates to identify the antibiotic concentration at which the initial inoculum is decreased by over 10^{-3}. The MBC value does not usually exceed several dilutions below the MIC which it closely parallels, and the MBC values do not correspond in magnitude with much elevated concentrations required to kill biofilms of the same isolate. Standards for the MIC and MBC assays are managed by the Clinical & Laboratory Standards Institute (CLSI). The MIC and MBC assays, as outlined by the CLSI guidelines, are considered gold standard for antibiotic screening; adherence to these tests is required by the Food and Drug Administration, which regulates antibiotics for clinical use. As such, researchers and drug manufactures use the MIC and to a lesser extent MBC as the exclusion criteria for antibiotic candidates for clinical use. There are no regulatory guidelines or clinical standards which require the use of biofilms for testing potential clinical compounds despite their role in infectious disease.

Conventional microbiology remains fixated on the actively dividing planktonic phenotype, completely ignoring the biofilm and its central role in the most devastating, debilitating, and costly infections observed in healthcare today. Bacteria in biofilms display antibiotic tolerance many orders of magnitude higher than the MIC and MBC values; this has already been discussed in detail above. Furthermore, the MIC values do not always predict the relative order of efficacy observed in analogous benchtop tests using biofilms. For example, Fig. 2 compares the antibiotic kill profiles and MICs for two clinical gold-standard antibiotics, vancomycin and gentamicin, against an experimental compound designated CZ-01127 produced by Curza Global, LLC (Salt Lake City, UT). CZ-01127 is a first-in-class tri-alkylnorspermidine-biaryl antibiotic [65]. All three of the antibiotics shown in Fig. 2 have an MIC of 1 μg/ml against the S. *aureus* ATCC 6538 isolate; yet each antibiotic has marked differences when tested against biofilms of the same isolate. For example, at 500 μg/ml vancomycin was no different from biofilm controls, yet CZ-01127 reduced the bacterial numbers in the biofilm to below the detection threshold of the experiment [65]. Undoubtedly, useful antibiofilm compounds have been overlooked in laboratory notebooks and drug libraries of companies and academic researchers throughout the years of devout fixation on planktonic phenotypes as the gold standard for antibiotic screening.

Corresponding clinical evidence supports the use of biofilms for testing antibiotic compounds. People with cystic fibrosis (CF) are prone to pulmonary exacerbations characterized by *P. aeruginosa* biofilm lung infections. In 2003, a study was conducted to determine the relationship between the MICs of *P. aeruginosa* isolates collected from the biofilms in the sputum of 262 patients with cystic fibrosis and the patients' response to parenteral antibiotics. The authors summarized this study: "When treatment outcomes were plotted against tobramycin or ceftazimide MICs of patients' least susceptible or most predominant *P. aeruginosa* isolates, there was no obvious correlation between the isolate antibiotic susceptibility and the patient response." [66] A subsequent retrospective analysis in 2008 on 110 patients who received antibiotics for exacerbations compared their treatment outcomes to the susceptibility of the cultured isolate based on both the MIC and outcomes and assays from the Calgary Biofilm Device. The authors concluded that, "Patients in this study who were treated with antibiotics that inhibited biofilm growth of at least one sputum isolate showed significant improvement in sputum bacterial density and length of hospital stay." [67] Researchers looking to repurpose the anticancer drug cisplatin for treating *P. aeruginosa* infections found, "The MIC of cisplatin and tobramycin against planktonic *P. aeruginosa* cells were 6.25 μM and 2.65 μM, respectively. However, tobramycin could not kill the biofilm cells at 2 × MIC…while cisplatin was able to kill substantial amount of biofilm cells with nearly 100 times reduction of *P. aeruginosa* biofilm cells." They tested cisplatin in a murine model of corneal infection, finding that, "Cisplatin showed efficient killing capacity on *P. aeruginosa* cells from infected mouse corneas and . . . a significant reduction in the bacterial loads from cisplatin treated corneas as compared to control corneas." [68] Taken together these studies motivate the need for adopting biofilms in antimicrobial testing protocols, not to supplant the MIC and MBC assays, but to supplement these invaluable gold-standard tests.

Preventing Biofilm Infection

Even in the most aseptic procedures, there is an almost certainty that microbes will gain access to the surgical site (see discussed above). This motivates prophylactic strategies for preventing opportunistic bacteria from progressing toward advanced implant-related biofilm infection. Preventing infectious bacteria from colonizing biomaterials is by necessity the focus of clinical efforts because simple methods for clearing established biofilms are limited at best. These infections frequently require costly surgical intervention, removal or replacement of implants, extensive debridement, delayed wound closure, and long courses of clinical antibiotics; they are associated with great cost both financially to the healthcare system and physically to patients in terms of high morbidity (e.g., amputations and loss of function) and high mortality rates. The first several hours after tissue inoculation, either through traumatic wounding or during a surgical procedure, are critical for infection prophylaxis.

Clinicians and researchers realized early in the development of many of the first-line antibiotics that antimicrobial chemotherapy does not always control infection, even if the isolates are deemed susceptible through laboratory screening techniques and biomaterials are not introduced in the site [69]. Quantitative bacteriology studies on animals with experimental lesions from the mid-1900s elucidate the process of infection progression in mammalian tissues [69–71]. Within the first 4 h of inoculation with planktonic bacteria above the minimum infectious dose, the host defenses eliminate most of the bioburden, decreasing it by ~10^6 of the initial inoculate; ensuing local infection is determined by the small number of bacteria surviving the early offensive by host defenses. The mechanisms underlying the initial kill and subsequent rebound of bacterial numbers is not completely understood but is likely an interplay of several factors. Contaminants might arrive at the site in a metabolically less active lag phase of growth; progression into a log-phase mode of growth might account for the observed rebound. Furthermore, the bacteria surviving the initial onslaught of host defenses might form biofilm aggregates, effectively shrouding themselves from host defenses thereby creating a nidus for subsequent infection. Indeed biofilms have been documented to form in the absence of biomaterials and are proven in several types of infection [72, 73].

Administration of systemic antibiotics is most effective when used in the perioperative timeframe preceding the rapid rebound of bacterial numbers (1–6 h). The authors who first described this phenomenon in 1961 concluded, "There is a definite short period when the developing staphylococcal dermal or incisional infection may be suppressed by antibiotics. This effective period begins the moment bacteria gain access to the tissue and is over in 3 hours. Systemic antibiotics have no effect on the primary staphylococcal infections if the bacteria creating the infection have been in the tissue longer than 3 hours before the antibiotics are given. Antibiotics cause maximum suppression of infection if given before bacteria gain access to tissues." [69] The emerging antibiotic recalcitrance of these rebounding bacteria to compounds to which they were previously susceptible just hours before indeed suggests phenotypic changes like those observed in antibiotic-tolerant biofilms. This early timeframe of antibiotic efficacy was first called "the effective period of preventive antibiotic action" but now is more simply "the decisive period." This work forms the basis for current clinical guidelines directing the use of prophylactic systemic antibiotics for surgical procedures. The importance of the decisive period was affirmed by a large clinical study following the infection outcomes of 1708 patients who received prophylactic antibiotics; the study authors concluded that, "The lowest rate of surgical-wound infection occurred in the patients who received antibiotics from 0 to 2 hours before surgery. The trend toward higher rates of infection with each successive hour that antibiotic administration was delayed after the surgical incision was significant." [74]

On this theoretical basis, several noteworthy technologies, providing brief local antibiotic fields around the implant within the decisive period, have gained clinical acceptance. In a simplistic approach surgeons have resorted to the off-label use of vancomycin or tobramycin by peppering surgical sites and implants with pure antibiotic powders for prophylaxis. Antibiotic concentrations in local tissues are short-lived yet peak concentrations coincide strongly with the decisive period as shown in the blue trace in Fig. 3 displaying the concentration profile of vancomycin from aspirates of the joint capsule of a total knee arthroplasty treated with intraoperative antibiotic powder [75]. These approaches have been shown in multiple studies to decrease the risk of orthopedic surgical site infection especially in spinal procedures [76, 77]. The authors of a large meta-analysis on the topic concluded that "local administration of vancomycin powder appears to be associated with significantly lower risk of SSIs [surgical site infections], deep incisional SSIs, and *S. aureus* SSIs." [76] In a more sophisticated approach, the pacemaker manufacturer Medtronic

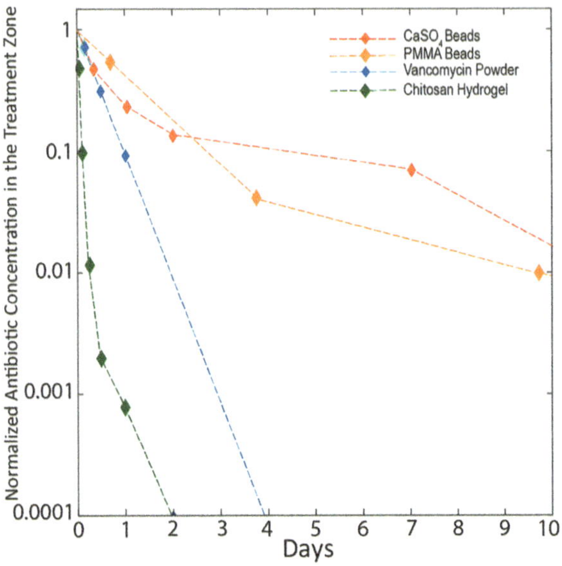

Fig. 3 Representative release curves of common local antibiotic delivery products and techniques including antibiotic loaded bone cements of $CaSO_4$ (red) and PMMA (orange), vancomycin powder (blue), and a representitive hydrogel (green). Release curves were adapted from literature references

produced an absorbable pacemaker envelope loaded with minocycline and rifampin to prevent postimplantation biofilm colonization. Multiple independent studies have shown this device to both decrease infection rates and remain cost-effective [78, 79]. Unfortunately, despite the effectiveness of this technology, it would be near impossible to get similar products to market using the same regulatory strategy because of changes in the current FDA regulatory climate. This is unfortunate for the many patients who suffer debilitating and even lethal infections, which could otherwise be prevented.

Strategies for Treating Chronic Biofilm Infections

Established infections on the other hand might benefit from very high, sustained local doses of antimicrobials, concentrations which cannot be safely achieved systemically due to toxicity of the most susceptible tissues (hepato-, nefro-, ototoxicity, etc.). Several off-label local-delivery approaches have gained clinical acceptance yet remain limited in practice. Bone cements are routinely loaded with antibiotics, formed into small beads, and packed within a surgical site. Antibiotic beads can be made from both polymethylmethacrylate (PMMA) or calcium sulfate ($CaS0_4$) cements. PMMA curing is exothermic, limiting application to heat-stable antibiotics. Tight and rapid polymerization results in antibiotic entrapment; only 3–7% of antibiotic is ever released, and peak concentrations are low (5–25 μg/ml); this might be an effective prophylactic strategy yet remains ineffective at resolving biofilm-related infections of orthopedic implants or devitalized bone as occurs in osteomyelitis [80, 81]. $CaSO_4$ cements accommodate heat-sensitive antibiotics, are biodegradable, and completely release the antibiotic payload. $CaS0_4$ beads can achieve local antibiotic concentrations of ~10,000 μg/ml, 10,000x the MIC, and have shown to be an effective adjuvant to standard surgical debridement to resolve device-related infections and osteomyelitis [82]. Yet dissolution of $CaSO_4$ causes drainage at the implant site and increases risk of heterotopic ossification [83]. Antibiotic $CaS0_4$ beads have shown efficacy against planktonic bacteria but limited impact against biofilms [83–85]. Numerous experimental technologies for local administration of antibiotics have been discussed, prototyped, and tested. Most of these technologies comprise preloaded, nonrefillable, polymer-based delivery systems: antimicrobial coatings, hydrogels, sponges, and degradable plastics.

The most critical limitation of these off-label and experimental approaches is their inability to sustain high antibiotic concentrations at the surgical site. These systems use a rapidly depleted internal antibiotic reservoir; the diffusive driving force and drug release swiftly diminish as the internal reservoir of antibiotic is depleted. For example, with $CaS0_4$ beads, antibiotic concentrations in tissues can diminish by over 90% within just 5 days [86]; the antibiotic concentrations of loaded chitosan hydrogels for orthopedic applications diminished over 99.9% within one day (Fig. 3), [87] and even sophisticated hydrogels, like those with which we have experience [88], show rapidly diminishing postimplantation release profiles. Nonrefillable sustained release systems are further constrained by the poor stability of many clinically relevant antibiotics in hydrated environments at body temperature. Penicillins, cephalosporins, carbapenems, quinolones, and some glycopeptides rapidly degrade at 37 °C in DI-H_2O, decreasing by as much as 50% within 7 days [89]. These degradation rates are further affected by host enzymes when moved from ideal DI-H_2O conditions to the tissues of the body. The local-releasing technologies discussed are not obsolete but may be relegated to prophylactic applications as discussed in the preceding section, applications which require strong burst kinetics within the decisive period of the perioperative window. There is an evident need for medical devices which can sustain high antibiotic concentrations at the implant site. This will inevitably come with the challenges of managing a cumbersome regulatory system, which is not amenable to combination products lacking clear predicate devices.

Planktonic Inocula in Animal Models of Infection

There is currently no viable replacement to study infection processes or determine infectious doses than animal models of infection. There, likewise, has never been a greater need for relevant models of infection as infectious diseases are more common and concerning than ever. Malaria and tuberculosis continue to debilitate millions across the globe each year. Lyme disease is steadily climbing in the USA over the past two decades. And as medical device usage becomes commonplace in medical practice throughout the world, the risk of biofilm device-related infection grows in tandem.

In vitro susceptibility assays primarily hinge on outcome measures related to planktonic bacteria. So it is with animal models of infection; infectious processes are primarily based on outcome measures following inoculation with planktonic bacteria. Bacterial cultures are typically grown in broth media or on agar surfaces, adjusted to a specific concentration and inoculated (typically in log phase growth) into excision wounds, surgical sites, ocular regions, peritoneal spaces, intramuscular tissues, or joint capsules among other anatomical locations. In orthopedic or other device applications, bacteria are typically inoculated on or near the material of interest. The common denominator between all of these methods is the use of suspended, free-floating, planktonic bacteria that are in a highly metabolically active state.

Inoculation with planktonic bacteria typically leads to acute infection. These can quickly become raging and require euthanasia of an animal within days following inoculation/surgery [90]. With lower doses, i.e., when less bacteria are inoculated initially, infection can be slower to develop and more closely model a chronic infection that has a delayed onset. Planktonic contaminants can form biofilms over time and lead to more chronic-type infection outcomes, yet time to infection and type of infection (e.g., latent versus acute) may vary [91].

Animal models of infection have been foundational to the development of therapies and protocols that protect mankind from infectious processes, and this section is not intended to demean the value that has been and will continue to be obtained from planktonic bacteria as initial inocula. Healthcare challenges simply continue to be broader than the scope of medical application and knowledge. There is still ground to cover. As we consider the decades-long persistence of biofilm-related infections in healthcare settings including military medical facilities, and the lack of therapeutics that specifically target biofilm-dwelling organisms, it is evident that advancements need to be made and additional model development is necessary. In this context, at least five limitations exist when using planktonic bacteria as initial inocula:

1. It is estimated that >99% of bacteria in natural ecosystems (including human skin, see chapter "We Begin to Target the Biofilm") dwell in the biofilm phenotype. Laboratory conditions are primarily optimized for log phase growth of planktonic cells in broth or agar systems. These conditions may not accurately reflect natural environments, growth states, substrates, or contamination mechanisms.
2. Planktonic cells are more readily cleared by immune system components than cells residing in a biofilm. In video presentations at the 8th ASM Conference on Biofilms, Brian Pettygrove (of Dr. Phil Stewart's lab) demonstrated the ability of neutrophils to readily phagocytose bacterial cells in the planktonic state, but after just 4 h of immature biofilm/aggregate development, neutrophil phagocytosis became frustrated and failed to control the clusters effectively. If inoculated in the planktonic state, host immunity may gain the upper hand before infection sets in, which is consistent with the lack of reproducible infection outcomes in many animal models of planktonic-derived infection, such as models of osteomyelitis.
3. Planktonic bacteria are more susceptible to antibiotics than bacteria in biofilms. If prophylactic or extended courses of antibiotic are administered in animal models, data collection may be skewed as the contaminants may be eradicated before infection sets in.
4. Depending on the method of inoculation, planktonic bacteria may spread/diffuse through host tissue or fluids more rapidly than bacteria in a biofilm. This could dilute the concentration of bacteria per area, potentially allowing the host to handle the bioburden more easily and/or reduce attachment to an intended device that has also been implanted.
5. Due to their high metabolic state, infection signals caused by planktonic bacteria may be exaggerated, acute, and less chronic in nature. Whether this were advantageous or detrimental would be application-dependent but potentially problematic if biofilm-related infection is desired to be modeled.

Biofilms as Initial Inocula

The current clinical climate warrants additional model development and therapeutic optimization toward biofilm-related infections. As a specific example, 5-year mortality rates for patients who suffer from periprosthetic joint infections (considered to be complicated by biofilms) are now higher than mortality rates of breast cancer, melanoma, Hodgkin's lymphoma, and other cancers [92, 93]. As total joint replacement procedures are estimated to rise drastically in the coming decades, infection numbers and mortality risks will likewise increase, that is, if no changes are implemented in treatment modalities or therapeutic approaches. As we identify and target the problems that underpin these outcomes, we may influence these numbers for the better and, more importantly, alleviate suffering in those who are affected.

One strategy to improve clinical treatments for biofilm-related infections is to specifically develop technologies and therapeutics that target biofilms. This can begin by confirming that infections are, in fact, influenced by the presence of biofilms. One component of this approach could consist of using biofilms as initial inocula in animal models of infection to more closely mimic bacteria in their natural state of existence. As discussed in chapter "We Begin to Target the Biofilm", normal flora organisms in human skin dwell in the biofilm phenotype and may serve as initial contaminants to surgical sites at the time of surgery. Models, therapeutics, and protocols that take this into account may improve biofilm-related wound and/or surgical site infections. Exogenous biofilm contaminants also have the potential to contaminate wound sites and lead to unique infections that are immediately recalcitrant to antibiotic therapy. For example, wounds that a soldier may suffer on the battlefield can be immediately contaminated with biofilm-dwelling organisms as bacteria in natural ecosystems (e.g., soil) preferentially dwell in biofilms. Similarly, traumatic injuries such as motorcycle accidents, falls that occur during hiking, automobile accidents, sports injuries, and many others are susceptible to biofilm contamination at the point of injury [94–96]. Our group has developed animal models that use biofilms as initial inocula to mimic these situations to improve therapies that treat and/or prevent infections specifically compromised by biofilm contaminants. Pilot testing in sheep indicates that a 48 h course of dual prophylactic antibiotics (gentamicin + cefazolin) is ineffective at reducing biofilm inocula any more than the host itself (publication pending), supporting the need for biofilm-specific therapies and/or protocols.

More recently, we have developed a sheep model of heterotopic ossification (HO)—a pathology involving ectopic bone formation that affects 60% or more of wounded warriors who suffer from blast-related trauma in current conflicts [97]. Previous work in rodents has indicated that the presence of staphylococcal isolates exacerbates the formation of HO [98]. We advanced this work to large animal models and further included the use of established biofilms as initial inocula. Biofilms were grown on the surface of glass beads (Fig. 4), placed into a traumatically injured wound site with bone chips, periosteal disruption, wound VAC therapy, and a simulated IED blast. Outcomes have indicated that biofilm contaminants lead to a

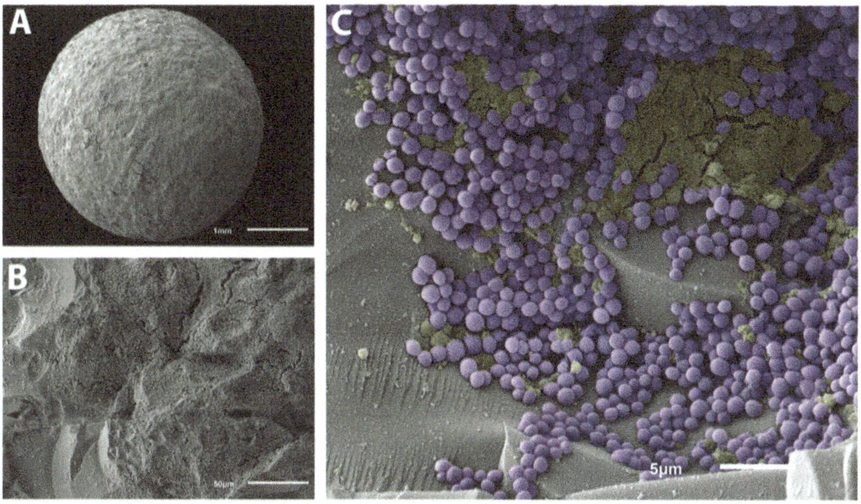

Fig. 4 Scanning electron micrographs of silica beads (sanded briefly to roughen the surface) and biofilms of *S. aureus* ATCC 6538. Silica (sand) with established biofilms may contaminate the wound site of an injured service member during a traumatic blast or other injury. (**a**) An ~3 mm silica bead. (**b**) Topography of a silica bead with biofilms of *S. aureus* ATCC 6538 grown on the surface. (**c**) Higher power image with false coloring that indicates early biofilm formation (purple cells) and EPS production (yellow matrix) on a silica bead surface (gray background)

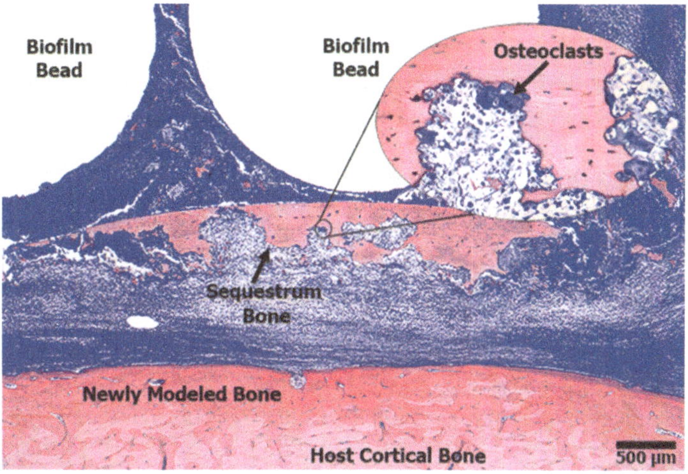

Fig. 5 Histological section from a sheep model of HO. Biofilms on silica beads were used as initial inocula. Bone sequestra (hallmark of osteomyelitis) with osteoclast activity and significant inflammatory response developed near the biofilms. In this case, no antibiotics were required as the infection progressed slowly in chronic fashion without fever or other significant distress to the animal. Host bone modeling in the cortical region indicates an active response to the bone trauma. Data are pending publication

chronic, low grade-type infection with sequestra formation (Fig. 5), and, in this particular model, exacerbated HO formation. This work adds to the arsenal of models available to determine the effect of biofilm contaminants on the development of biofilm-related infection and pathological processes.

Multiple differences can be considered by using biofilms as initial inocula as opposed to planktonic bacteria:

1. Bacteria in biofilms more closely model the dwelling state of bacteria in natural ecosystems.
2. Biofilms are more efficient at evading host immune defenses than planktonic cells.
3. Inoculation with established biofilms can immediately provide recalcitrance to antibiotic therapies, thus more closely modeling clinical scenarios that suffer from biofilm-related infection.
4. Bacteria in mature biofilms are less metabolically active than planktonic bacteria, and thus have improved chance of developing low-lying, chronic states of infection.

Conclusions

The biofilm underpins the etiology of the most difficult to treat infections observed in the clinic today. Most of these infections are caused by otherwise commensal organisms but our reliance on artificial materials for use in implanted prostheses and medical devices has shifted this delicate balance toward microbe pathogenesis as these artificial materials serve as a nidus for biofilm formation. We must consider the biofilm as we develop the clinical standards and animal models that will take medicine into the future.

Disclaimer Author DLW has financial interest in Curza Global, LLC.

References

1. Mangram, A. J., et al. (1999). Guideline for prevention of surgical sit infection, 1999. *Infection Control and Hospital Epidemiology, 20*, 247.
2. https://journals.lww.com/jbjsjournal/Fulltext/2006/08000/Surgical_Site_Infection_in_the_Elderly_Following.3.aspx?__hstc=215929672.82af9c9a98fa600b1bb630f9cde2cb5f.1535155200301.1535155200302.1535155200303.1&__hssc=215929672.1.1535155200304&__hsfp=2025384311
3. Cogen, A. L., Nizet, V., & Gallo, R. L. (2008). Skin microbiota : A source of disease or defence ? *The British Journal of Dermatology, 158*, 442–455.
4. Mortazavi, S. M. J., Schwartzenberger, J., Austin, M. S., Purtill, J. J., & Parvizi, J. (2010). Revision total knee arthroplasty infection: Incidence and predictors. *Clinical Orthopaedics and Related Research, 468*, 2052–2059.

5. Moriarty, T. F., et al. (2016). Orthopaedic device-related infection: Current and future interventions for improved prevention and treatment. *EFORT Open Reviews, 1*(13–17), 89.
6. Gahlot, R., Nigam, C., Kumar, V., Yadav, G., & Anupurba, S. (2014). Catheter-related bloodstream infections. *International Journal of Critical Illness and Injury Science, 4*(162–167), 161.
7. Franco, J. J., & Ruisi, P. (2014). Infectious complications of percutaneous cardiac procedures. *Interventional Cardiology, 6*, 445–452.
8. Achermann, Y., Goldstein, E. J. C., Coenye, T., & Shirtliff, M. E. (2014). Propionibacterium acnes : From commensal to opportunistic biofilm- associated implant pathogen. *Clinical Microbiology Reviews, 27*, 419–440.
9. Grice, E. A., et al. (2009). Topographical and temporal diversity of the human skin microbiome. *Science, 324*, 1190–1192.
10. Portillo, M. E., Corvec, S., Borens, O., & Trampuz, A. (2013). Propionibacterium acnes : An underestimated pathogen in implant-associated infections. *BioMed Research International, 2013*, 1.
11. Sahin, Y. A. F., & Schwyzer, H. K. (2013). Characteristics and outcome of 16 periprosthetic shoulder joint infections. *Infection, 41*, 613–620.
12. Levy, P. Y., et al. (2008). Propionibacterium acnes postoperative shoulder arthritis : An emerging clinical entity. *Clinical Infectious Diseases, 46*, 1884–1886.
13. Contracture, C., et al. (2009). Pilot study of association of bacteria on breast implants with capsular contracture. *Journal of Clinical Microbiology, 47*, 1333–1337.
14. Sampedro, M. F., et al. (2010). A biofilm approach to detect bacteria on removed spinal implants. *Spine (Phila. Pa. 1976), 35*, 1218–1224.
15. Nichols, R. L. (1984). Prevention of infection in high risk gastrointestinal surgery. *The American Journal of Medicine, 76*, 111–119.
16. Collaborative, G. (2018). Surgical site infection after gastrointestinal surgery in high-income, middle-income, and low-income countries: A prospective, international, multicentre cohort study. *The Lancet Infectious Diseases, 18*, 516–525.
17. Lachiewicz, M. P., Moulton, L. J., & Jaiyeoba, O. (2015). Pelvic surgical site infections in gynecologic surgery. *Infectious Diseases in Obstetrics and Gynecology, 2015*, 1–8.
18. Nablo, B. J., Rothrock, A. R., & Schoenfisch, M. H. (2005). Nitric oxide-releasing sol – gels as antibacterial coatings for orthopedic implants. *Journal of Biomaterials, 26*, 917–924.
19. Hetrick, E. M., Schoenfisch, M. H., Hetrick, E. M., & Hetrick, E. M. (2006). Reducing implant-related infections: Active release strategies. *Chemical Society Reviews, 35*, 780–789.
20. Mclaughlin, J. S., Hornick, R. B., Ronica, M., & Frank, M. (2019). Sources of contamination in open heart surgery. *JAMA, 230*, 1415–1418.
21. Chu, N. S., Chan-myers, H., Aam, R. M., Ghazanfari, N., & Antonoplos, P. (1999). Levels of naturally occurring microorganisms on surgical instruments after clinical use and after washing. *American Journal of Infection Control, 27*, 315–319.
22. Agarwal, A. A. (2019). University of T. of O. Intraoperative guard to avoid biofilm and infection at pedicle screw & bone interface. In *Poster presentation at the international combined orthopaedic research societies annual meeting*.
23. Sender, R., Fuchs, S., & Milo, R. (2016). Revised estimates for the number of human and bacteria cells in the body. *PLoS Biology, 14*, e1002533.
24. Hanazaki, K., Shingu, K., Adachi, W., Miyazaki, T., & Amano, J. (1999). Chlorhexidine dressing for reduction in microbial colonization of the skin with central venous catheters: A prospective randomized controlled trial. *The Journal of Hospital Infection, 42*, 165.
25. Mann, T. J., Orlikowski, C. E., Gurrin, L. C., & Keil, A. D. (2001). The effect of the biopatch, a chlorhexidine impregnated dressing, on bacterial colonization of epidural catheter exit sites. *Anaesthesia and Intensive Care, 29*, 600–603.
26. Krizek, T. J., & Robson, M. C. (1975). Evolution of quantitative bacteriology in wound management. *American Journal of Surgery, 130*, 579–584.
27. Robson, M. C., Lea, C. E., Dalton, J. B., & Heggers, J. P. (1968). Quantitative bacteriology and delayed wound colsure. *Surgical Forum, 19*, 501–502.

28. Robson, M. C., & Heggers, J. P. (1969). Bacterial quantification of open wounds. *Plastic and Reconstructive Surgery, 44*, 19–24.
29. Manley, G. (2014). Shigella: A highly virulent elusive pathogen. *Current Tropical Medicine Reports, 1*, 81–87.
30. Greig, J. D., Todd, E. C. D., Bartleson, C., & Michaels, B. (2010). Infective doses and pathogen carriage. In *Food safety education conference* (pp. 1–31).
31. Leggett, H. C., Cornwallis, C. K., & West, S. A. (2012). Mechanisms of pathogenesis, infective dose and virulence in human parasites. *PLoS Pathogens, 8*(10–12), e1002512.
32. Evans, D. G., Miles, A. A., & Niven, J. S. F. (1948). The enhancement of bacterial infections by adrenaline. *British Journal of Experimental Pathology, 29*, 20–39.
33. Miles, A. A. & Niven, J. S. F. The enhancement of infection during shock pro- DUCED by bacterial toxins and other agents. (1949).
34. Kass, E. H. (1956). Diagnostic urology asymptomatic infections of the urinary tract. *Trans-American Physiology, 69*, 56–64.
35. Antiseptic, U. (1957). Bacteriuria and the diagnosis of infections of the urinary tract. *A.M.A. Archives of Internal Medicine, 100*, 709–714.
36. Robson, M. C., & Heggers, J. P. (1970). Delayed wound closures based on bacterial counts. *Journal of Surgical Oncology, 2*, 379–383.
37. Liedberg, N. C. F., Reiss, E., & Artz, C. P. (1955). The effect of bacteria on the take of split-thickness skin grafts in rabbits. *Annals of Surgery, 142*, 92–96.
38. Bowler, P. G. (2003). The 10(5) bacterial growth guideline : Reassessing its clinical relevance in wound healing. *Ostomy/Wound Management, 49*, 44–53.
39. Elek, S. D., & Conen, P. E. (1957). The virulence of staphylococcus pyogenes for man; a study of the problems of wound infection. *British Journal of Experimental Pathology, 38*, 573–586.
40. James, R. C., & MacLeod, C. J. (1961). Induction of staphyloccocal infections in mice with small inocula introduced on sutures. *British Journal of Experimental Pathology, 42*, 266–277.
41. Ratner, B. D. (2004). *Biomaterials science: An introduction to materials in medicine*. Amsterdam: Academic Press.
42. Costerton, J. W., Geesey, G. G., & Cheng, K.-J. (1978). How bacteria stick. *Scientific American, 238*, 86–95.
43. Segal, A. W. (2005). How neutrophils kill microbes. *Biological Membranes, 23*, 197–223.
44. Nizet, V. (2007). Understanding how leading bacterial pathogens subvert innate immunity to reveal novel therapeutic targets. *The Journal of Allergy and Clinical Immunology, 120*, 13–22.
45. Sarantis, H., & Grinstein, S. (2012). Subversion of phagocytosis for pathogen survival. *Cell Host & Microbe, 12*, 419–431.
46. Baltimore, R. S., & Mitchell, M. (1980). Immunologic investigations of mucoid strains of Pseudomonas aeruginosa : Comparison of susceptibility to opsonic antibody in mucoid and nonmucoid strains. *The Journal of Infectious Diseases, 141*, 238–247.
47. Stewart, p. S. (2014). Biophysics of biofilm infection. *Pathogens and Disease, 70*, 212–218.
48. Matsui, H., et al. (2005). Reduced three-dimensional motility in dehydrated airway mucus prevents neutrophil capture and killing bacteria on airway epithelial surfaces. *Journal of Immunology, 175*, 1090–1099.
49. Parkhurst, M. R., & Saltzman, W. M. (1992). Quantification of human neutrophil motility in three-dimensional collagen gels. Effect of collagen concentration. *Biophysical Journal, 61*, 306–315.
50. Parkhurst, M. R., & Saltzman, W. M. (1994). Leukocytes migrate through three-dimensional gels of midcycle cervical mucus. *Cellular Immunology, 156*, 77–94.
51. Hänsch, G. M., Brenner-Weiss, G., Prior, B., Wagner, C., & Obst, U. (2008). The extracellular polymer substance of Pseudomonas aeruginosa: Too slippery for neutrophils to migrate on? *The International Journal of Artificial Organs, 31*, 796–803.
52. Leid, J. G. (2009). Bacterial biofilms resist key host defenses once in biofilms, bacterial pathogens resist antibiotics and withstand several host-defense measures, including phagocytosis. *Microbe, 4*, 66–70.

53. Herant, M. (2006). Mechanics of neutrophil phagocytosis: Experiments and quantitative models. *Journal of Cell Science, 119*, 1903–1913.
54. Gutiérrez, D., Hidalgo-Cantabrana, C., Rodríguez, A., García, P., & Ruas-Madiedo, P. (2016). Monitoring in real time the formation and removal of biofilms from clinical related pathogens using an impedance-based technology. *PLoS One, 11*(1–17), e0163966.
55. Clardy, J., Fischbach, M. A., & Currie, C. R. (2009). The natural history of antibiotics. *Current Biology, 19*, R437–R441.
56. Costerton, J. W., Cheng, K. J., Geesey, G. G., Ladd, T. I., Nickel, J. C., Dasgupta, M., & Marrie, T. J. (1987). Bacterial Biofilms in Nature and Disease. *Annual Review of Microbiology, 41*, 435–464.
57. Jacques, M., Marrie, T. J., & Costerton, J. W. (1987). Review: Microbial colonization of prostehtic devices. *Microbial Ecology, 13*, 173–191.
58. Stewart, P. S., et al. (2016). Reaction-diffusion theory explains hypoxia and heterogeneous growth within microbial biofilms associated with chronic infections. *npj Biofilms Microbiomes, 2*, 1–8.
59. Ochman, H. G. L., & Found, G. (2002). Evolution of bacterial pathogenesis and Symbiosis. *Science, 292*, 1096–1099.
60. Stewart, P. S. (2015). Antimicrobial Tolerance in Biofilms. *Microbiology Spectrum, 3*, 1–30.
61. Wolcott, R. D., et al. (2010). Biofilm maturity studies indicate sharp debridement opens a time-dependent therapeutic window. *Journal of Wound Care, 19*, 320–328.
62. Khoury, A. E., Lam, K., Ellis, E., & Costerton, J. W. (1992). Prevention and control of bacterial infections associated with medical devices. *ASAIO Journal, 38*, 174–178.
63. Tan, S. Y., & Tatsumura, Y. (2015). Alexander Fleming (1881–1955): Discoverer of penicillin. *Singapore Medical Journal, 56*, 366–367.
64. Fleming, A. (1929). On antibacterial action of culture of Penicillium, with special reference to their use in isola- tion of B. influenzae. *British Journal of Experimental Pathology, 10*, 226–236.
65. Ashton, N. N., et al. (2019). In vitro testing of a first-in-class tri-alkylnorspermidine-biaryl antibiotic in an anti-biofilm silicone coating. *Acta Biomaterialia, 93*, 25–35. https://doi.org/10.1016/j.actbio.2019.02.010.
66. Smith, A. L., Fiel, S. B., Mayer-Hamblett, N., Ramsey, B., & Burns, J. L. (2003). Susceptibility testing of Pseudomonas aeruginosa isolates and clinical response to parenteral antibiotic administration: Lack of association in cystic fibrosis. *Chest, 123*, 1495–1502.
67. Keays, T., et al. (2009). A retrospective analysis of biofilm antibiotic susceptibility testing: A better predictor of clinical response in cystic fibrosis exacerbations. *Journal of Cystic Fibrosis, 8*, 122–127.
68. Yuan, M., et al. (2018). Repurposing the anticancer drug cisplatin with the aim of developing novel Pseudomonas aeruginosa infection control agents. *Beilstein Journal of Organic Chemistry, 14*, 3059–3069.
69. Burke, J. F. (1961). The effective period of preventive antibiotic action in experimental incisions and dermal lesions. *Surgery, 50*, 161–168.
70. Miles, A. A., Burke, J., & Miles, E. M. (1957). The value and duration of defence reactions of the skin to the primary lodgement of bacteria. *British Journal of Experimental Pathology, 38*, 79–96.
71. Polk, H. C., & Milest, A. A. (1973). The decisive period in the primary infection of muscle by escherichia coli. *British Journal of Experimental Pathology, 54*, 99–109.
72. Gristina, A. G., Oga, M., Webb, L. X., & Hobgood, C. D. (1985). Adherent bacterial colonization in the pathogenesis of osteomyelitis. *Science, 228*, 990–993.
73. Omar, A., Wright, J., Schultz, G., Burrell, R., & Nadworny, P. (2017). Microbial biofilms and chronic wounds. *Microorganisms, 5*, 9.
74. Classen, D. C., et al. (2015). The timing of prophylactic administration of antibiotics and the risk of surgical-wound infection. *The New England Journal of Medicine, 326*, 281–286.

75. Johnson, J. D., et al. (2017). Serum and wound vancomycin levels after Intrawound administration in primary total joint arthroplasty. *The Journal of Arthroplasty, 32*, 924–928.
76. Chiang, H. Y., Herwaldt, L. A., Blevins, A. E., Cho, E., & Schweizer, M. L. (2014). Effectiveness of local vancomycin powder to decrease surgical site infections: A meta-analysis. *The Spine Journal, 14*, 397–407.
77. Molinari, R. W., Khera, O. A., & Molinari, W. J., III. (2012). Prophylactic intraoperative powdered vancomycin and postoperative deep spinal wound infection: 1,512 consecutive surgical cases over a 6-year period. *European Spine Journal, 21*, 476–482.
78. Kay, G., et al. (2018). Cost-effectiveness of TYRX absorbable antibacterial envelope for prevention of cardiovascular implantable electronic device infection. *Journal of Medical Economics, 21*, 294–300.
79. Bloom, H. L., et al. (2011). Implantation success and infection in cardiovascular implantable electronic device procedures utilizing an antibacterial envelope. *Pacing and Clinical Electrophysiology, 34*, 133–142.
80. Moojen, D. J. F., et al. (2008). In vitro release of antibiotics from commercial PMMA beads and articulating hip spacers. *The Journal of Arthroplasty, 23*, 1152–1156.
81. Anagnostakos, K., et al. (2008). Antimicrobial properties and elution kinetics of linezolid-loaded hip spacers in vitro. *Journal of Biomedical Materials Research. Part B, Applied Biomaterials, 87*, 173–178.
82. McPherson, E., Dipane, M., & Sherif, S. (2013). Dissolvable antibiotic beads in treatment of periprosthetic joint infection and revision arthroplasty – the use of synthetic pure calcium sulfate (Stimulan®) impregnated with vancomycin & tobramycin. *Reconstructive Review*, 3.
83. Helgeson, M. D., Potter, B. K., Tucker, C. J., Frisch, H. M., & Shawen, S. B. (2009). Antibiotic-impregnated calcium sulfate use in combat-related open fractures. *Orthopedics, 32*, 323.
84. Howlin, R. P., et al. (2015). Antibiotic-loaded synthetic calcium sulfate beads for prevention of bacterial colonization and biofilm formation in periprosthetic infections. *Antimicrobial Agents and Chemotherapy, 59*, 111–120.
85. Trujillo, J. M., Logue, M. E., Kunkel, R., & Demas, C. P. (2017). Off-label usage of absorbable beads containing antibiotics for prevention of surgical site infections. *Wounds, 29*, E84–E87.
86. Aiken, S. S., Cooper, J. J., Florance, H., Robinson, M. T., & Michell, S. (2014). Local release of antibiotics for surgical site infection management using high-purity calcium sulfate: An in vitro elution study. *Surgical Infections, 16*, 54–61.
87. Rhodes, C. S., et al. (2017). Evaluation of a chitosan-polyethylene glycol paste as a local antibiotic delivery device. *World Journal of Orthopedics, 8*, 130.
88. Lane, D. D., Fessler, A. K., Goo, S., Williams, D. L., & Stewart, R. J. (2017). Sustained tobramycin release from polyphosphate double network hydrogels. *Acta Biomaterialia, 50*, 484–492.
89. Samara, E., et al. (2017). Antibiotic stability over six weeks in aqueous solution at body temperature with and without heat treatment that mimics the curing of bone cement. *The Bone & Joint Research, 6*, 296–306.
90. Sinclair, K. D., et al. (2013). Model development for determining the efficacy of a combination coating for the prevention of perioperative device related infections: A pilot study. *Journal of Biomedical Materials Research. Part B, Applied Biomaterials, 101*, 1143–1153.
91. Bernthal, N. M., et al. (2010). A mouse model of post-arthroplasty Staphylococcus aureus joint infection to evaluate in vivo the efficacy of antimicrobial implant coatings. *PLoS One, 5*, e12580.
92. Zmistowski, B., & Casper, D. S. (2013). Periprosthetic Joint Infection Increases the Risk of one-year mortality. *The Journal of Bone and Joint Surgery. American Volume, 95*, 2177–2184.
93. Rezapoor, M., & Parvizi, J. (2015). Prevention of periprosthetic joint infection. *The Bone & Joint Journal, 99B*, 3–10.
94. Williams, D. L., & Costerton, J. W. (2012). Using biofilms as initial inocula in animal models of biofilm-related infections. *Journal of Biomedical Materials Research. Part B, Applied Biomaterials, 100*(B), 1163–1169.

95. Williams, D. L., et al. (2012). Experimental model of biofilm implant-related osteomyelitis to test combination biomaterials using biofilms as initial inocula. *Journal of Biomedical Materials Research. Part A, 100 A*, 1888–1900.
96. Williams, D. L., et al. (2012). In vivo efficacy of a silicone-cationic steroid antimicrobial coating to prevent implant-related infection. *Biomaterials, 33*, 8641–8656.
97. Isaacson, B. M., Brown, A. A., Brunker, L. B., Higgins, T. F., & Bloebaum, R. D. (2011). Clarifying the structure and bone mineral content of heterotopic ossification. *The Journal of Surgical Research, 167*, e163–e170.
98. Pavey, G. J., et al. (2015). Bioburden increases heterotopic ossification formation in an established rat model. *Clinical Orthopaedics and Related Research, 473*, 2840–2847.

Index

D. L. Williams (ed.), *Targeting Biofilms in Translational Research, Device Development, and Industrial Sectors*,
https://doi.org/10.1007/978-3-030-30667-0

The manufacturer's authorised representative in the EU is Springer Nature Customer Service Centre GmbH, Europaplatz 3, 69115 Heidelberg, Germany. If you have any concerns regarding our products, please contact ProductSafety@springernature.com

Printed and bound by CPI Group (UK) Ltd, Croydon, CR0 4YY
30/09/2026
02225665-0011